A Fresh Concept of Software-resemblant Hardware to Leap to 6G and Future Networks

Micro/Nanotechnologies as Enablers of Pervasivity

Published 2024 by River Publishers

River Publishers

Alsbjergvej 10, 9260 Gistrup, Denmark

www.riverpublishers.com

Distributed exclusively by Routledge

605 Third Avenue, New York, NY 10017, USA

4 Park Square, Milton Park, Abingdon, Oxon OX14 4RN

A Fresh Concept of Software-resemblant Hardware to Leap to 6G and Future Networks / by Jacopo Iannacci.

Routledge is an imprint of the Taylor & Francis Group, an informa business

ISBN 978-87-7004-079-2 (paperback)

ISBN 978-10-4008-524-0 (online)

ISBN 978-1-003-49492-8 (ebook master)

A Publication in the River Publishers Series in Rapids

A Fresh Concept of Software-resemblant Hardware to Leap to 6G and Future Networks

Micro/Nanotechnologies as Enablers of Pervasivity

Jacopo Iannacci

Fondazione Bruno Kessler (FBK), Italy

Routledge
Taylor & Francis Group

NEW YORK AND LONDON

Dedication

To Brando, Pietro (Jr.), Moira, Rossana, Pietro (Sr.); to "Er Branco" and, in general terms, to all the recollections of lightheartedness

"I think, as soon as you cross that line where you are worrying about what other people say about you, you are in big trouble, because then you are always guessing. For me, I never really cared that much what people thought. Of course, a lot of people liked it; there have been other situations people haven't like it. In a way, both are irrelevant to me. Because – believe me – every night, when I go home, I know how well I played compared to how well I wanna play."

Pat Metheny – Interviewed during the "We live here" tour in Japan (1995)

"I'll put it simple.
If you're going hard enough left, you'll find yourself turning right.
(Turn right to go left.)"

The Fabolous Hudson Hornet – From the Walt Disney Pictures movie "Cars" (2006)

"Not everyone can become a great artist.
But a great artist can come from anywhere."

Anton Ego – From the Walt Disney Pictures movie "Ratatouille" (2007)

Contents

Preface

In the last decade, we witnessed an unprecedented spread of services and applications on the move, supported by the relentless evolution of telecommunication systems, technologies and standards. To date, less than 30 years after the commercial diffusion of cellular phones (under the umbrella of 2G), for a large part of the society in developed countries, it seems inconceivable to have a mobile handset just for placing voice calls and exchanging short text messages.

All of this was possible thanks to the evolution of hardware (HW) technologies, especially in the fields of electronics and semiconductors, along with that of software (SW), although the main co-design approach to the development of HW–SW devices, systems and infrastructure, remained unaltered across the past half a century.

Looking ahead, the emerging massive paradigms of 6G, future networks (FN), super-internet of things (IoT) and tactile internet (TI), are going to pose unprecedented challenges that cannot be properly addressed by traditional approaches, despite still being effective today for 5G. This is because 6G and FN will not simply be a *faster 5G*.

Artificial intelligence (AI) will be the cornerstone of future telecommunication standards. On one hand, AI will support the evolution of functionalities offered to the end-user (service plane), in continuity with what is already ongoing today. From a different perspective, and in full disruption to the current state, AI will be massively leveraged in the management of the network infrastructure itself (network operation plane). The set of resulting ramifications is incredibly large and, to a certain extent, still unexplored and unpredictable.

Unprecedented concepts, like, e.g., self-evolution, self-sustenance and high-resilience of small portions of the network, are forecasted to be pivots of 6G and FN, to optimize operation and ensure maintenance of key performance indicators (KPI), at any time and in all conditions. This will lead to the so-called systems of systems, posing challenges in terms of orchestration, to ensure local flexibility, while keeping the entire network working as a whole.

In addition, data storage and computation capacities will scatter down to any tinier branch of the network (at the edge), boosting the concept of edge intelligence (EI). This will set challenges in terms of low-cost, low-complexity, low-power and highly miniaturized HW implementing EI, of power availability and provision to ensure operation, along with connectivity to the rest of the network, i.e., from the edge to the core, and so on.

In a nutshell, 6G and FN will have to leverage a broad set of HW and SW key enabling technologies (KET), addressing, among other items, ubiquity of services and coverage, operation in frequency bands as high as sub-THz (up to 300 GHz) and above (toward THz), along with optical signals, including visible light communications (VLC), etc. Bearing all this in mind, it is straightforward that unprecedented approaches, along with thinking out of the box, are necessary to conceive, design, develop and deploy the network of the future.

This book takes the first steps in bringing disruption to the design and development approaches applied to HW–SW systems and sub-systems for future 6G, FN and super-IoT. To this end, it is highlighted that the typical development stream of telecommunication standards and services has been following a top-down direction, i.e., from high-level KPI, to protocols, standards, SW technologies, and, only after, to HW infrastructure, systems, sub-systems, and, eventually, physical components. This work makes an effort to overturn such a paradigm, proposing and supporting a bottom-up stream, which starts from very-low complexity HW components, like sensors, actuators and transducers.

The target is not to dismiss currently in use design approaches in favor of the novel ones here described, but rather to complement them, aiming to reinforce exploitation of all the possibilities and degrees of freedom (DoF) offered by the identified KET, regardless of whether they are HW or SW technologies. To this end, the bottom-up approach, from basic HW sensors and transducers, to HW–SW systems, infrastructure and services, is potentially a keystone.

In the classical design strategies in use, HW technologies for basic components are involved towards the end of the development process, when the physical infrastructure is already set. This means that the interaction between the *system* and HW technologies is *one way* (i.e., top-down), and takes shape in a set of specifications that must be satisfied by the sensor or transducer. Diversely, the bottom-up approach supported by this work carries, among other things, the benefit of involving HW technologies for low-complexity devices in the discussion, already in the early stages of a new paradigm conception. This results in having at disposal all the DoF offered by HW technologies, typically made invisible by the classical top-down approach.

The work reports practical examples of services and functionalities, especially at the edge of the network, which can be empowered and, in other cases, reinvented by exploiting to the full the DoF of HW technologies. Microtechnologies and nanotechnologies, intended with a broad meaning, which includes devices, systems (MEMS/NEMS) and materials, are identified as a crucial asset to trigger a new field of convergence, in which miniaturized HW components can be merged with AI technologies, helping to draw the paint of 6G and FN, still unknown to a relevant extent. In addition, the new paradigm of quantum computing is embodied in the discussion around the emerging quantum technologies. The impact of the unprecedented approaches discussed in this work is expected to be so significant that a reformulation of the same concept of HW is offered.

In conclusion, the challenges of 6G and FN will bring more disruption than what we have witnessed in the last 40 years, in the long haul from the first generation of mobile communications to the current 5G. This means that disruptive approaches at the conceptual, prior that at technology, level, will be necessary to turn the current visions into reality. Within such a context, this book brings valuable elements into the discussion, which will be certainly useful in the continuation of the journey towards 6G and FN.

Melike Erol-Kantarci, Ph.D.

Canada Research Chair (Tier 2) in AI-Enabled Next-Generation Wireless Networks and Full Professor, School of Electrical Engineering and Computer Science, University of Ottawa.

Jacopo Iannacci, Ph.D.

Researcher in MEMS technologies at Fondazione Bruno Kessler (FBK), Italy.

About the Author

Jacopo Iannacci (Senior Member of the IEEE) was born in Bologna, Italy in 1977. He received his MSc (Laurea) degree in electronics engineering in 2003 and his PhD in information and telecommunications technology in 2007, both from the University of Bologna, Italy. He received Habilitation as Full Professor in Electronics from the Italian Ministry of University and Research (MUR), in 2021.

He worked in 2005 and 2006 as visiting researcher at the DIMES Technology Center (currently Else Kooi Lab) of the Technical University of Delft, the Netherlands, focusing on packaging and integration of RF-MEMS (radio frequency passives in MEMS technology). In 2016, he visited as seconded researcher the Fraunhofer Institute for Reliability and Microintegration IZM in Berlin, Germany, to conduct high-frequency characterization of RF-MEMS components. Since 2007, he has been a researcher (permanent staff) at the Center for Sensors & Devices of Fondazione Bruno Kessler in Trento, Italy.

His research interests and experience fall in the areas of finite element method (FEM) multi-physics modelling, compact (analytical) modeling, design, optimization, integration, packaging, experimental characterization and testing for reliability of MEMS and RF-MEMS devices and networks for sensors and actuators, energy harvesting (EH-MEMS) and telecommunication systems, with applications in the fields of 5G, Internet of Things (IoT), as well as future 6G, Tactile Internet (TI) and Super-IoT.

Dr. Iannacci has authored about 150 scientific contributions. He is currently associate editor of Springer Microsystem Technologies, and series editor of the IEEE Communications Magazine (ComMag). He has been involved in several international conferences as symposium chair/co-chair, session chair, technical program committee member, international advisory board member, tutorial lecturer and invited speaker.

List of Abbreviations

1G	1st generation of mobile telecommunications
2G	2nd generation of mobile telecommunications
2.5G	1st evolution of 2G
3F	Forecasted Functional Features
3G	3rd generation of mobile telecommunications
3.5G	1st evolution of 3G
3.75G	2nd evolution of 3G
4G	4th generation of mobile telecommunications
5G	5th generation of mobile telecommunications
6G	6th generation of mobile telecommunications
ADC	Analog-to-Digital Converter
AI	Artificial Intelligence
AM	Additive Manufacturing
ASIC	Application Specific Integrated Circuits
B5G	Beyond-5G
BOx	Buried Oxide
BS	Base Station
BTC	Brain-Type Communications
CDMA	Code-Division Multiple Access
CIM	Computing-In-Memory
CMOS	Complementary Metal-Oxide Semiconductor
CMR	Critical Mega-Requirements
CNT	Carbon Nanotubes
CRAS	Connected Robotics and Autonomous Systems
CvE	Core vs. Edge model
DC	Direct Current
DL	Downlink
DLa	Device Layer
DoF	Degree of Freedom
E2E	End-To-End

EB	Exabyte
EH	Energy Harvester/Harvesting
EH-MEMS	EH realized in MEMS technology
EI	Edge Intelligence
EMBB	Enhanced Mobile Broadband
EMI	Electromagnetic Interference
FDMA	Frequency-Division Multiple Access
FinFET	Fin Field-Effect Transistor
FLC	Four-Leaf Clover
FN	Future Networks
Ga	Gallium
GaAs	Gallium Arsenide
GaInN	Gallium Indium Nitride
Ge	Germanium
HLa	Handle Layer
HW	Hardware
I2D	Identifying-Defining-Deriving model
IC	Integrated Circuits
IMT	International Mobile Telecommunications
IMU	Inertial Measurement Unit
In	Indium
IoE	Internet of Everything
IoT	Internet of Things
ITU	International Telecommunication Union
KET	Key Enabling Technologies
KPI	Key Performance Indicators
LCD	Liquid-Crystal Display
LSB	Least Significant Bit
LTE	Long Term Evolution
M2M	Machine-To-Machine
MEMS	MicroElectroMechanical-Systems
MIMO	Multiple Input Multiple Output
mMIMO	Massive MIMO
mm-wave(s)	Millimeter wave(s)
MMTC	Massive Machine-Type Communications
MSB	Most Significant Bit
MTC	Machine-Type Communications
NEMS	NanoElectroMechanical-Systems
NFV	Network Function Virtualization
OFDMA	Orthogonal Frequency-Division Multiple Access
PAt	Power Attenuator
PC	Personal Computer

PCB	Printed Circuit Board
PCU	Passive Control Unit
PECVD	Plasma-Enhanced Chemical Vapor Deposition
PhS	Phase Shifter
QoS	Quality of Service
QT	Quantum Technologies
RF	Radio Frequency
RFFE	Radio Frequency Front End
RF-MEMS	RF passive components in MEMS technology
Rx	Reception
SAEWII	Space, Air, Earth, and Water Integrated Infrastructure
SDN	Software-Defined Network/Networking
SEM	Scanning Electron Microscope/Microscopy
SHeaLDS	Self-Healing Light guides for Dynamic Sensing
SiC	Silicon Carbide
SiLCHI	System in a Low Complexity Hardware Item
SiP	System-In-Package
SMS	Short Message Service
SoC	System-On-Chip
SOI	Silicon On insulator
Super-IoT	Super Internet of Things
SW	Software
SwU	Switching Unit
TDMA	Time-Division Multiple Access
TI	Tactile Internet
TMT	Technology Mega-Trends
TRM	Transmission/Reception (Tx/Rx) Module
TWV	Through-Wafer Vias
Tx	Transmission
UAV	Unmanned Air Vehicles
UL	Uplink
URLLC	Ultra Reliable and Low Latency Communications
USB	Universal Serial Bus
USD	United State Dollar
V2V	Vehicle-To-Vehicle
V2X	Vehicle-To-Everything
VLC	Visible Light Communications
W2W	Wafer-To-Wafer
WEAF Mnecosystem	Water, Earth, Air and Fire Micro/Nanotechnologies Ecosystem
WLP	Wafer-Level Packaging

WPT	Wireless Power Transfer
WSN	Wireless Sensor Network
XR	Extended Reality
ZB	Zettabyte

Introduction

With the uptake of 4G (4th generation of mobile telecommunications) more than one decade ago, we got accustomed to a relentless increase of data and services on the move. The deployment of 5G (5th generation of mobile telecommunications) is seamlessly advancing crucial key performance indicators (KPI), such as average data rate per user, to break the 1 Gbps frontier, and latency, at <5 ms. This will keep improving quality of service (QoS) and pervasivity, along with a more immersive end-user experience. Setting the horizon further ahead, to 2030 and later, 6G (6th generation of mobile telecommunications) will take KPIs to numbers 100–1000 times better than what 5G will achieve in the next few years.

Bearing this in mind, a continuum across recent and future generations is easily identifiable. Yet, a rough underrating would be thinking the transition from one standard to another as a mere improvement of KPI. In fact, 4G-LTE (long term evolution) triggered the novel concept of network function virtualization (NFV), which is the "softwarization" of portions of the physical network. The resulting increase of abstraction makes the network functioning more adaptable and reconfigurable. These concepts will be further advanced by 5G, while 6G is expected to bring unprecedented disruption at various levels.

Artificial intelligence (AI) will be exploited in a threadlike fashion, at any level of the network physical infrastructure. This will introduce up to date unknown features, like self-sustaining, self-evolution and high-resilience of small portions of the infrastructure, pioneering the concept of a *network of networks*. Each segment of the infrastructure will bear a high degree of independence, while working at the same time as a whole, in full orchestration with the rest of the network.

Given such a scenario, this book claims that the established and currently in use paradigms for the design and development of hardware–software (HW-SW) systems, are not appropriate to address the challenges of 6G and, further ahead, of future networks (FN). In response, unprecedented design approaches are here suggested, relying on a fresh reinterpretation of the standard concept of HW, with specific attention to the network edge.

Before stepping into the main statements of this work in a more detailed fashion, a few considerations are needed on how telecommunications evolved from 1G (1st generation of mobile telecommunications) in the late-1970s to date, along with networked services.

The starting scenario is that of local nodes, mainly personal computers (PC) – individual/corporate use – and mainframes – mostly corporate use, locally provided with their own applications, computational and memory capacities, which were exchanging limited amounts of data with remote units, mainly over landlines. With the commercialization of the internet, around mid-1990s, a relentless trend to centralization (to the cloud) of services, computation and data was triggered.

In parallel, mobile services remained limited to voice and text messaging, roughly until the transition to 3G (3rd generation of mobile telecommunications), in the first years of the 2000s. Later on, end-users became familiar with accessing the internet, remote data and contents (e.g., video streaming) on the move. More recently, conceptually introduced by 4G-LTE and in progress with 5G, mobile standards started to evolve towards an integrated platform for massive data exchange. This means that, beside services for end-users, like voice, video, social media access, etc., machine-type communications (MTC) will increasingly rely on 5G, 6G, etc. infrastructure and services.

Putting together the mentioned trends related to the evolution of services provided by the network (network services plane), along with the AI-boosted concept of a network of networks (network operation plane), the need for disruption in the design approaches for HW–SW systems unquestionably emerges.

Bearing in mind the complex and intricate scenario depicted above, this work develops some conceptual tools that may help address the resulting technical challenges, with particular focus on the 6G and FN network edge. Within the mentioned HW reconceptualization, a pivotal role is forecasted for microtechnologies and nanotechnologies, intended with a broad meaning, which embraces, amongst others, devices, systems and materials.

Going into more detail, the book content is arranged as follows. After a brief introduction on fundamental concepts related to mobile services and networks, the first effort in Chapter 1 is that of offering a simplified and comparative description of the wireless communication standards, from 1G

(1st generation of mobile telecommunications), back in the early-1980s, to the current 5G and, again, to the future 6G and FN. This is done by developing a graphical-based descriptive approach, called the core vs. edge (CvE) model. The CvE scheme highlights how extensively important features of each generation, e.g., services, data storage/elaboration and physical infrastructure, are leaning toward the core or the periphery (edge) of the network. The other objective of Chapter 1 is to frame the broad field of microtechnologies and nanotechnologies (MEMS/NEMS). To this end, rather than reporting a state-of-the-art review on devices and solutions realized in such technologies, a different approach is pursued. First, a brief discussion is developed on the evolution of micro/nanotechnologies through time, highlighting the overlap with that of standard semiconductors. Subsequently, the fundamental steps of two pillar manufacturing approaches to the realization of micro- and nano-devices, namely, surface and bulk micromachining, are reported. This is done to help the non-expert readers familiarize themselves with the basics of MEMS and NEMS. All in all, multiple examples and references of innovative sensors, devices and systems successfully implemented in the micro/nano-world will be frequently mentioned in the rest of the book.

Moving on, Chapter 2 makes an effort to provide motivation to the need for a reformulation of the concept of HW, in view of 6G and FN, and of their functioning. This undertaking builds upon the development of a conceptual framework, which is also provided with a graphical 3D implementation, easing the understanding. The ad-hoc constructed tool is named the identifying-defining-deriving (I2D) model, and is based on a deductive approach. The I2D features three sets of elements, which can also be visualized as dimensions in a three-dimensional space. The first is that of critical mega-requirements (CMR), which are high-level needs, and KPI, increasingly important for the transition to 6G/FN. In this work, seven CMR are identified, and just to mention a couple of them, edge intelligence (EI) and energy distribution are covered. The second dimension of the I2D model is that of technology mega-trends (TMT). These are technology trends that will be driven by CMRs, i.e., when moving from high-level KPI towards the actual physical (HW–SW) implementation and deployment of the network. The set of TMT is composed of 11 entries, among which the increase of miniaturization/integration, decrease of power consumption and incorporation of intelligence at low-complexity device level, are reported. Finally, the third (vertical) dimension of the I2D model is that of forecasted functional features (3F). Taking a further step down, i.e., towards the physical realization of components, 3F are actual functionalities that, putting together the demands outlined by CMR and TMT, future low-complexity HW devices will be expected to satisfy. The set of 3F features seven entries, among which multi-functional basic/monolithic HW devices and self-reacting sensors/actuators are reported. The third dimension of 3F closes the loop with microtechnologies and nanotechnologies. In fact, per 3F, examples taken from the current literature are reported on devices, systems and solutions that satisfy, at least partially, the identified demands. As a last item, Chapter 2 reports a set of correlation matrices. In particular, per each of the seven 3F, the correlation with the bi-dimensional crossing of seven CMR and 11 TMT are graphically visualized.

After the background developed by the I2D model, the final Chapter 3 addresses the actual reformulation of the concept of HW and its formalization. First, the concept of a *HW–SW divide* is introduced, highlighting the cornerstone approaches to the development of HW–SW systems, which have remained unaltered for about five decades of telecommunications evolution, and that can turn into a limiting factor when facing the challenges ahead of 6G/FN. Subsequently, the concepts of *separation* and *symmetry*, and in particular their increase, are discussed and reported as a solution to overcome the limitations of the HW–SW divide. Starting from such a base of knowledge, a conceptual framework for the development of future micro-/nano-devices with unprecedented characteristics (as demanded from 6G/FN) is introduced. It is named the WEAF Mnecosystem (water, earth, air and fire micro/nanotechnologies ecosystem), and builds an analogy between the features of low-complexity HW components and the four elements in nature. In particular, earth and air represent the classical concepts of HW and SW, respectively. Diversely, water addresses simple devices, like sensors and transduces, with diversified functionalities and abilities, among which stand those of self-reacting and self-repairing. Fire, eventually, addresses micro- and nano-devices dealing with energy, i.e., able to harvest energy from the environment, store and transfer it, thus providing power where needed, when needed, in all the tiniest ramifications of the network edge. Chapter 3 also develops a rather extensive discussion of examples in the literature that report MEMS/NEMS devices, technologies and solutions, complying, at least in part, with the requirements and features formally introduced by the WEAF Mnecosystem.

From 6G to MEMS/NEMS Physical Transducers – A Free Fall in Complexity

Abstract

The aim of this first chapter is that of providing the necessary base of knowledge for the easy understanding of the subsequent discussion, to be developed across the following two chapters. With this view, coverage is provided in two distinct fields of scientific knowledge. First, a simplified and comparative description of the wireless communication standards, from 1G, back in the early-1980s, to the current 5G and, again, to the upcoming 6G and FN, is offered. This is done by developing a graphical-based descriptive approach, called the CvE model. The CvE scheme highlights how extensively important features of each generation, e.g., services, data storage/elaboration and physical infrastructure, are leaning toward the core or the periphery (edge) of the network. The other target is to frame the broad field of microtechnologies and nanotechnologies (MEMS/NEMS). To this end, a brief discussion is developed on the evolution of micro/nanotechnologies through time, highlighting the overlap with that of standard semiconductors. Subsequently, the fundamental steps of two pillar manufacturing approaches to the realization of micro- and nano-devices, namely, surface and bulk micromachining, are reported.

1.1 An Overview of Telecommunication Standards Evolution

Looking back over nearly half a century of advancements, the world of telecommunications has marked unthinkable steps ahead, from the perspective of employed technologies, provided services and achieved performance. Yet, some cornerstone concepts remained unaltered across such a wide timespan. If one compares the functionalities offered by 1G in the early-1980s with those of 5G, the two standards seem to have nothing in common, apart from the letter "G". Nevertheless, the concept of service coverage split into cells, and handover from one another (i.e., cellular communications) introduced for the first time by 1G, is still the pillar of current standards [1], [2]. The rationale is schematized in Figure 1.1, where the transmitting mobile device ("S" – speaker) located in cell A, communicates seamlessly with the receiving one ("L" – listener), while the latter moves from cell B to cell C [3].

Figure 1.1: Working principle of cellular communications, with different radio towers covering adjacent portions of space (i.e., cells), thus ensuring continuity of services during handover.

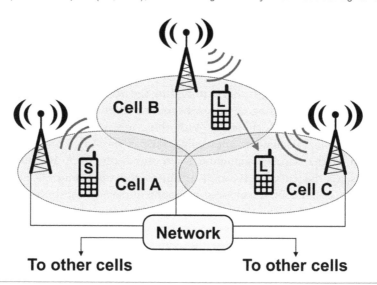

Talking about the improvement and proliferation of services and functionalities available on the move, the synergy between the advancement of semiconductor technologies and electronics, along with the development of unprecedented and more efficient techniques to code and transmit data, played a crucial role. To this end, the fully analogue channel splitting of 1G, named FDMA, was scarcely efficient, both in terms of security of broadcasted data, and of usage of the allocated bandwidth [3]. The subsequent digital techniques of splitting channels over time (TDMA) and of coding data (CDMA), both introduced under 2G and further developed later on, improved significantly the efficiency in sending and receiving data with respect to several performance indicators, including power consumption. More recently, further evolutions in data coding and splitting over the available spectrum, like OFDMA [3], introduced under 3.75G and widely exploited by 4G-LTE, pushed KPI further ahead.

A thorough description of the technologies and solutions employed by each generation is not going to be covered here. This is because reference books and scientific literature already exist. Also, such a discussion would fall outside the scope of the book. That said, a brief recap of the progressing of KPI across different generations is provided in Tables 1.1 and 1.2. The summary is split in two tables, which are Table 1.1, grouping 1G to 4G-LTE, and Table 1.2, focusing on 5G and beyond.

Table 1.1: Summary of the most relevant characteristics of mobile communications standards, from 1G to 4G-LTE [4].

Generation	Data rate	Band	Bandwidth	Services
1G	2.4 kbps	800 MHz	30 kHz	Voice
2G	10 kbps	850–900 MHz	200 kHz to 1.25 MHz	Voice; data
2.5G	50–200 kbps	1.8–1.9 GHz	200 kHz 200 kHz	Voice; data
3G	384 kbps	800–900 MHz	5 MHz 1.25 MHz	Voice; data; video calling
3.5G	5–30 Mbps	1.8–2.1 GHz	5 MHz 1.25 MHz	Voice; data; video calling
3.75G	100–200 Mbps	1.8–5.8 GHz	1.4–20 MHz	Voice; data; video calling; online gaming; high definition television
4G-LTE	3 Gbps – DL[a] 1.5 Gbps – UL[b]	1.8–3.5 GHz	1.4–20 MHz	Voice; data; video calling; online gaming; high definition television

[a]DL – Downlink
[b]UL – Uplink

Starting from a merely quantitative analysis, the transition from 1G to the current 5G and beyond, marks a relentless increase of KPI, above which the average data rate is certainly relevant. In particular, an abrupt step was

Table 1.2: Summary (essential) of pivotal KPI in the transition from 5G to 6G [5].

	5G	Beyond 5G (B5G)	6G
Reference devices	Smartphones Wearables Drones	Smartphones Wearables Drones XR devices	CRAS[a] XR[b] and BTC[c] devices Body implants
Data rate	1 Gbps	100 Gbps	1 Tbps
E2E[d] delay requirement	5 ms	1 ms	below 1 ms
Radio delay requirement	100 ns	100 ns	10 ns
Processing delay	100 ns	50 ns	10 ns
E2E reliability requirement	99.999 %	99.9999 %	99.99999 %
Frequency bands	Sub-6 GHz mm-wave (fixed access)	Sub-6 GHz mm-wave (fixed access)	Sub-6 GHz mm-wave (mobile access) Sub-THz (up to 300 GHz) Non-RF (optical, VLC[e], etc.)

[a]CRAS – Connected robotics and autonomous systems
[b]XR – Extended reality
[c]BTC – Brain-type communication
[d]E2E – End-to-end
[e]VLC – Visible light communication

introduced by 4G-LTE, as the data rate passed from the order of Mbps to Gbps. Given such a new reference value, 5G will consolidate the Gbps per user, while 6G is expected to mark a further leap to the Tbps range.

Further interesting aspects can be grasped looking at data in a qualitative fashion, and this leads to the reason for splitting the pre-5G and 5G-and-beyond eras in two. Focusing on the services implemented from 1G to 4G in Table 1.1, the generations before 3G were exclusively providing voice calls and simple data (i.e., SMS). 3G introduced video calls, while subsequent evolutions of 3G brought internet access on the move, along with video streaming, online services, etc., until the smartphone and social media access era, triggered by 4G-LTE, in the early 2010s.

The context is radically different when looking at Table 1.2, referring to 5G and beyond. Leaving aside KPI for a while, it is relevant to focus on the rows reporting the reference devices and the frequency bands. Already today, in the first phase of 5G rollout, apart from smartphones, we are used to additional devices connected to the network, like smartwatches and wearables.

Moreover, the appearance of drones, intended as mobile networks nodes, has to be highlighted. Stepping to B5G and 6G, unprecedented devices pop up into the set of items connected to the network. They are both directly related to end-users, as in the case of XR devices, body implants and BTC applications, as well as to machines communicating with other machines, i.e., CRAS.

The mentioned items offer an insight into the disruption that 5G and the subsequent FN will mark with respect to previous generations. If until 4G the network was mainly structured to handle end-user services (see rightmost column in Table 1.1), with 5G, data generated by M2M activities start traveling across the infrastructure. The implications in terms of increased volumes of data are huge. In fact, the term M2M embodies a broad variety of applications, including V2V and V2X communication, autonomous driving, industrial automation, CRAS, remote surgery, etc. Moreover, wide paradigms like the IoT, IoE and TI, based on the distribution to the network edge of sensing and computing functionalities, will weigh on the infrastructure.

Now, putting together the demands of M2M applications with those of XR and BTC, a few collective considerations can be drawn on Table 1.2. First, the relevant increase of average data rate from the Gbps to the Tbps range, from 5G to 6G, is solidly motivated. Then, the emerging relevance of KPI like E2E latency (delay) and reliability, not so crucial to previous generations, can be easily inferred, when dealing with applications like XR, autonomous driving, and remote surgery. As a result of such demands, the spectrum exploited to allocate data will be increasingly higher in frequency, venturing mm-waves, the sub-THz region, also surpassing the RF domain, with the exploitation of optical signals and VLC.

In summary, the unprecedented rationale developed by 5G discussed above, was effectively framed by the three main directions of EMBB, URLLC and MMTC, as established by ITU (IMT vision for 2020 and beyond) [6], [7], and shown in Figure 1.2.

The three corners define a triangular area in which 5G-empowered applications fall conceptually. Depending on the pivotal requirements a certain service needs, it is located in a specific portion of the space. For instance, autonomous driving has high-reliability and low-latency as crucial KPI. Therefore, it is placed close to the bottom-right corner in Figure 1.2. Diversely, 3D-video, holography and VR are located toward the top-corner because of their need to exchange large amounts of data over the network. Since its introduction back in 2015, the ITU scheme in Figure 1.2 is undergoing continuous refinement and extension [8–12]. Yet, it is employed by the main stakeholders who are developing 5G, and its rationale is likely to be a reference for 6G and FN.

Figure 1.2: The ITU 5G application space based on the EMBB, URLLC, MMTC triangle. (All the images and thumbnails used to compose the graphic are powered by Freepik.com.)

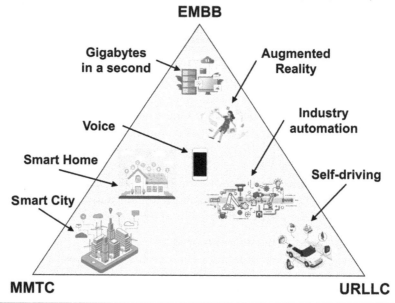

EMBB

Gigabytes in a second

Augmented Reality

Voice

Industry automation

Smart Home

Self-driving

Smart City

MMTC

URLLC

1.1.1 The core vs. edge (CvE) model to understand 6G and FN

What is discussed in the previous pages highlights that when looking into current and future telecommunication standards, one is confronting a high-complexity scenario. As such, B5G, 6G and FN, can be described from diverse perspectives, highlighting certain aspects and keys of interpretation, at the expense of others. The current scientific literature is rich with remarkable works that sketch perspectives and formulate plausible scenarios of 6G and FN, from the point of view of services/functionalities [13–16], KPI [17–20], network coverage and infrastructure [21], [22], along with sets of KET [[23]–[26], regarded as pivotal to enable the former items. The *fil rouge* in developing these studies is the adoption of a top-down approach, focused on a system- and application-level vision, in discussing both SW and HW technologies for 6G and FN. This is undoubtedly a sound approach, especially in scenarios still far from being standardized, like those of interest here.

The approach employed in this work is different, and to a certain extent unprecedented, if compared to those available in literature. For the benefit of recapping, the focus is aimed at HW technologies for low-complexity and miniaturized physical components, like sensors, actuators and transducers. To this end, microtechnologies and nanotechnologies (MEMS/NEMS) are identified as pivotal to support, empower and enable several features of 6G and FN, particularly at the edge of the network. What is proposed, in this sense, is a reformulation of the classical concept of HW, addressed in the following chapters.

To prepare the ground for such a later discussion, past and future telecommunication standards are going to be briefly reviewed according to a scheme that helps understand the network physical infrastructure complexity against the services offered by each specific generation at stake. The mentioned descriptor is named the core vs. edge (CvE) model, and weighs, among other features, the balance between centralization and distribution of key elements, like the HW infrastructure, volume of transferred data, data storage and elaboration, along with services implementation and availability.

The CvE model is split into three main sections, which are data, services and physical infrastructure. Each entry, regardless of the section, is described by a double-pointing arrow, toward edge and core. The arrow thickness indicates how extensively an entry develops in the direction of the network periphery or center. It must be stressed that the CvE description is merely qualitative. Therefore, the magnitudes suggested by the arrows must not be intended in absolute, but rather in relative terms, for what concerns the edge-to-core balance of a certain item. For example, the volume of transferred data over the network does not refer to the exact amount of TB traveling across the infrastructure, but to whether data are equally or unequally moving to the edge or the core.

In addition, it must be highlighted that the concept of core within the CvE model is broader with respect to the usual conception in mobile networks. In fact, here the core is not limited to the *deeper* segments of the mobile infrastructure, but includes the internet, as well. This means that a given service, like a social media, is exploited by the end-user at the edge when he/she accesses it by means, e.g., of an app installed on the smartphone. Diversely, the same service is used at the core when the end-user reaches it through, e.g., the web interface on a laptop cable connected to the internet. For the matter of completeness, within the CvE model, if the same laptop was connected to the internet via a smartphone acting as hotspot, the used services would count as delivered toward the edge of the network. Similarly, if a smartphone was linked to the internet via WiFi to a home or office router, in turn connected to the cable infrastructure, the services used by that smartphone would count as delivered towards the core of the network.

1.1.1.1 The CvE model of 1G

That said, the following CvE model description is not going to be replicated per each generation, as only those marking significant changes compared to the past, add value to the discussion of interest in this work. The CvE model for 1G is shown in Figure 1.3.

Figure 1.3: The CvE model of 1G.

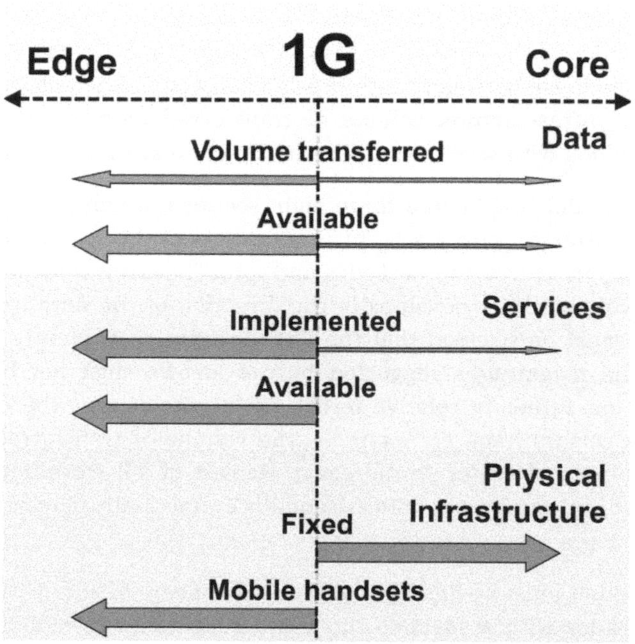

Looking at the data section, it is split into volume transferred, and available. The former is characterized by thin arrows, as the only data involved in 1G were voice conversations, mainly traveling in the vicinity of the network, i.e. among mobile handsets connected to cellular radio towers (fronthaul). On the other hand, the amount of data traveling toward the core, as well as the core itself, were scarcely developed, as the main purposes were those of orchestrating the whole infrastructure. Because of similar motivations, data were mainly available at the edge.

Concerning services (i.e., analogue voice conversations), they were implemented and available at the edge. The narrow stream of core implemented services refers to the collection of logs, mainly for network monitoring purposes (*back-end* services).

Stepping to the physical infrastructure section, the network was exclusively made of fixed items, from radio towers (fronthaul) to the backhaul and core. On the other hand, the *users* of services were exclusively mobile handsets, constituting the edge.

In light of the considerations on the CvE model in Figure 1.3, it can be preliminarily stated that 1G was a markedly distributed network.

1.1.1.2 The CvE model of 4G-LTE

The subsequent 2G and 3G are skipped, as they improved in a rather seamless fashion the CvE of 1G. Diversely, 4G-LTE introduced significant changes that deserve deeper insight. The CvE model of 4G-LTE is shown in Figure 1.4.

In the data section of 4G, both the volume and the availability of data mark a relevant increase toward the core. This is a clear indication of the shift to centralization, as opposed to previous generations. This can be inferred from the services section too, as services are mainly implemented towards the network core (i.e., in a centralized fashion), for being then made available at the edge. In practical terms, these considerations are linked to the dawn of the smartphone era, in conjunction with the widespread of internet access and services on the move.

Stepping to the physical infrastructure section of 4G, the landscape appears more complex than before. In fact, it features three subsections, named network end, human user end, and machine user end, plus two additional entries, named sensors and devices, and elaboration.

In an orderly manner, the network end is the classical terrestrial infrastructure. In comparison to previous generations, the fixed infrastructure went on growing toward the core, which means that standard radio towers (implementing cells) were increasingly deployed. What has to be highlighted is that under 4G a tiny arrow of the fixed infrastructure pops up toward the edge. With the proliferation of distributed services on the move, and the consequent increase of data traveling over the network, under 4G the idea of having smaller radio towers started to be investigated. Such access points cover limited portions of space, where data transfer is more intensive, e.g., in metropolitan and productive areas. This solution, experimented in the 4G era, is going to increase its importance under 5G, and will turn to be crucial in 6G and FN.

Figure 1.4: The CvE model of 4G-LTE.

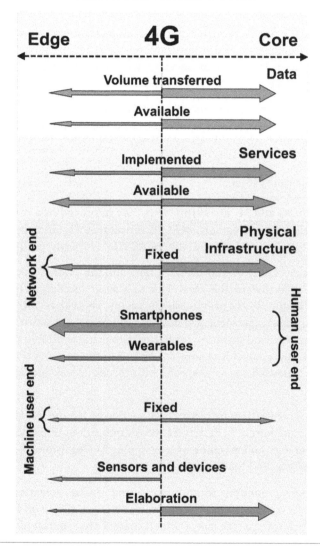

Then, what is referred to as the human user end, is the edge infrastructure constituted by the mobile terminal accessing remote services through 4G. As visible, such a subsection of the physical infrastructure marks a significant growth of smartphones (along with tablets, etc.), as well as the rising of wearable (smart) devices, directly connected to the network, like, e.g., smart watches, smart glasses, and so on. As already mentioned before, the possibility of smartphones (and other devices) to work as hotspots, allowing network access to other devices, makes them active parts of the edge infrastructure.

The reason why the previous infrastructure subsection needed to be further differentiated as *human*, has to do with another unprecedented concept introduced under 4G. In fact, the possibility to have part of the data generated from M2M applications, traveling on the same network used for voice calls and messaging chats, was introduced by this generation. Looking at the CvE scheme in Figure 1.4, this feature, identified as the machine user end, is marked by tiny arrows in both the edge and core directions. An actual transfer of M2M data toward the mobile network is not going to be achieved by 4G through its whole life. Yet, its inclusion and experimentation represent a remarkable disruption of 4G, which is going to acquire increasing centrality in 5G and beyond, also keeping in mind wide application paradigms like the IoT, IoE and TI.

Eventually, two other unprecedented entries appear under the physical infrastructure of 4G. The first one is named sensors and devices in Figure 1.4. This has to do with the proliferation of distributed sensing functionalities at the network edge. In particular, the IoT paradigm became very popular across the deployment of 4G after 2010. It brought on stage a plethora of different applications, like WSN, smart homes and smart factories, remote medicine (eHealth), etc., all linked by having distributed sensors acquiring data, to be then centrally processed, coming up with a feedback action to be locally implemented. Following the just discussed rationale of machine user end, 4G-LTE started to build, at least conceptually and with some preliminary experimentation, the scenario of embodying IoT-generated data in the mobile network.

Related to sensors proliferation and advancement of IoT concepts, the last entry of Elaboration must be discussed. In fact, the considerable amount of data acquired to the edge need to be collected and processed, leading to significant needs in terms of data storage and calculation capacities. In the 4G scenario, these resources are mainly located toward the network core, leading to a seamlessly increasing load of data to be transferred back and forth across the infrastructure. In this sense, 4G assumes the concept of centralization, previously triggered by 3G and its evolutions, and brings it to its maximum extent.

1.1.1.3 The CvE model of 5G

Changing paradigm, the CvE model of 5G is shown in Figure 1.5, and detailed in the following discussion.

Figure 1.5: The CvE model of 5G.

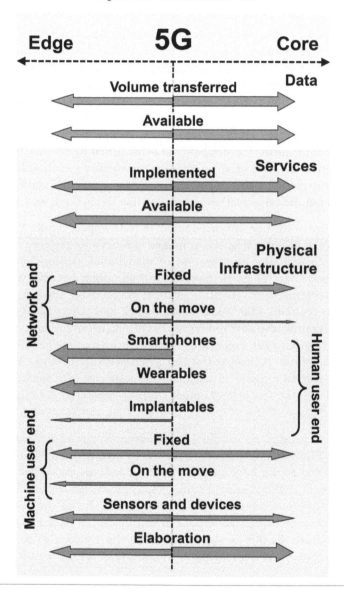

Wrapping together the most innovative aspects of 4G discussed above, 5G aims to implement them on a larger scale. Also, 5G is expected, through the whole timespan of its deployment, to start reversing the trend to centralization, pursued by 3G/4G for at least a couple of decades. Coming to the CvE model in

Figure 1.5 and to the data section, these considerations reflect in the increase both of the volume and availability toward the edge. Also, looking at the services section and starting from their availability, 5G (through its life) will bring the balance leaning toward the edge, as it is straightforward, given the massive increasing trend to access services on the move. Concerning implementation of services, the increase to the edge with respect to the core, is an important feature to be highlighted. In fact, this is an indication of the emerging trend towards the concept of a distributed (rather than centralized) network. The implications at infrastructural level are multiple and significant. One among the others, services implemented at the edge mean local availability of data storage and of elaboration capacities. The latter, along with the former inferences, build the cornerstone of this book, as will be discussed later on.

Moving to the physical segment of 5G, the network end is characterized by a growth of the fixed infrastructure tilted toward the edge. This is linked to the spread of access points covering smaller areas, yet providing increased data throughput in comparison to standard radio towers (as conceptually developed under 4G). To this end, the paradigm of small cells [27–30] is pivotal, and it will be further developed under 6G.

In full disruption with the past, the 5G network end introduces the concept of infrastructure on the move. This means that part of it will rely on moving access points, e.g., drones and airplanes [31], [32], rather than on standard fixed antenna systems. Despite looking almost like science fiction, the advantages of having a segment of the network relying on movable access points are invaluable in terms of reconfigurability and resilience. For instance, data throughput could be dynamically increased locally, where real-time demands emerge, e.g., where an emergency is ongoing (earthquakes, floods, etc.), or where large events (concerts, etc.) are taking place.

For what concerns the human user end, under 5G the ongoing increasing diffusion of smartphones will linger on, in conjunction with a significant spread-out of wearable devices. In addition, implantable devices will start to appear, too. Differently from wearables, which interact with human senses, like touch, hearing and vision, implantables will be able to directly interact with the human brain [33–35]. This will open up the novel field of BTC, in which the end-user interacts directly with services, without the need of physical actions, like pressing buttons or making gestures. Such implantable devices must not be envisaged as invasive surgical items, but rather as highly miniaturized sensors/transducers, to be placed over the skin (like a small band-aid), able to interact with human brain activity. This new paradigm is going to be conceptually introduced, and probably experimented on to a limited extent, under the umbrella of 5G. Nevertheless, it will be a pillar of 6G and FN.

Focusing now on the machine user end, previously introduced and discussed under 4G, the corresponding fixed infrastructure will grow both toward the edge and the core, as already mentioned. Notably, 5G is going to feature an on the move segment of such an infrastructure, mainly towards the edge. This is aligned with the above considerations around the network end of 5G.

Coming to the sensors and devices entry, 5G will increase their deployment to the edge, as suggested by IoT-like applications, previously covered. In disruption with the past, 5G will start employing such devices towards the core, as well. In brief, if sensors at the edge are mainly devoted to support and enrich end-user services, their purpose towards the core is more oriented to monitoring the network operation. Given the increased flexibility, reconfigurability and resilience of the network as a whole, addressed by 5G, more capillarization of sensing and controlling functionalities are to be expected.

Eventually, in light of the previous discussion, it is straightforward that data elaboration capacities will increase under 5G. If their growth at the core is predictable, it must be stressed that its increase to the edge is going to be proportionally relevant. This is a further indication of the growing trend to distribution, to be triggered by 5G and further developed afterwards.

1.1.1.4 The CvE model of 6G and FN

To conclude this section, a unique CvE model, both describing 6G and FN, is reported in Figure 1.6, as the main disruptions and founding paradigm shifts of 6G are expected to last for a long time. To this end, as before, several innovations will be conceptually introduced by a new generation, reaching full maturity in the following one. Though, the breaking of the rule of thumb, according to which each decade a new generation is introduced, and it reaches its plateau of deployment within the subsequent decade (15–20 years, from conception to next generation handover) is forecasted, starting from 6G. In fact, faster innovation cycles are likely to replace the previous ones, and already with 5G we are witnessing the dawn of such a trend.

The data and services sections can be analyzed as a whole and in a very brief way, as the center of gravity of both will heavily lean toward the edge. This provides an unmistakable indication of how markedly 6G and FN will be distributed networks.

Concerning the network end segment of the physical infrastructure, the slope is in favor of the edge. In particular, the infrastructure on the move will massively increase. To this end, the amount and diversification of movable items providing coverage everywhere will be relevant. An integrated heterogeneous

Figure 1.6: The CvE model of 6G and FN.

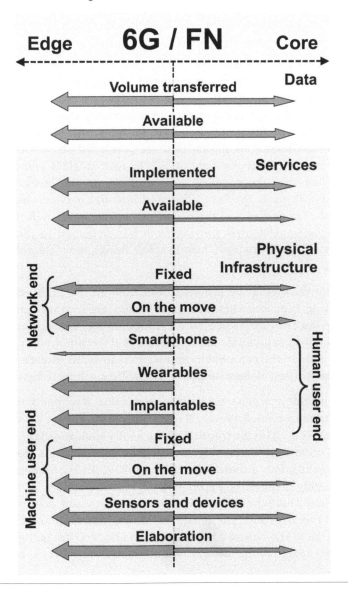

network, including drones, UAV, balloons, airplanes, satellites, along with on-water and under-water nodes across seas and oceans, is forecasted. Such a diversified infrastructure is named with the ad-hoc term SAEWII in [36], standing for space, air, earth, and water integrated infrastructure.

Moving to the human user end, a unique statement can summarize the whole discussion. Wearable and implantable devices will relentlessly increase, while the smartphone era will be nearing an end, probably in about one decade from now.

Concerning the machine user end, the massive increase of M2M communications, along with the widespread of IoT, IoE and TI application paradigms, will drive its growth toward the edge. As mentioned above, the largest increase will concern the on the move segment, which, differently from before, will mark a smaller, yet present proliferation towards the core as well. This is led by considerations similar to those developed above for the sensors and devices entry toward the core of 5G. In fact, if M2M communications at the edge lay onto the network service plane, on the other hand, their implementation towards the core is framed into the network service plane. In other words, the concepts of resilience, self-management, self-reaction and evolution, to be implemented starting from 6G, embodied in the *network of networks* term [37], will require, among other things, movable nodes towards its core.

Eventually, recapping the previous discussion on the proliferation to the edge of sensing functionalities, along with data elaboration capacities, the setting of the last two entries in Figure 1.6 does not require further discussion. The very last aspect to highlight is the reversal of elaboration, now clearly tilted to the edge, with all the consequent implications in terms of data storage and elaboration capacities, along with power availability, connectivity, and so on.

Before concluding, a general consideration is due. Recapping the rationale of the generations reviewed above, 1G was solidly sitting above the concept of a distributed network. With the transition from a fully analogue to the subsequent digital generations, the increasing tendency to centralization was triggered (after 2G), turning into a massive trend with 4G-LTE. 5G, through its whole life, today roughly in the *teenage years*, will unite centralization to an increasing decentralization. Thereafter, 6G and FN will invert the trend, mainly pointing to distribution of services and functionalities. Saying that, despite the complexity and articulation of the network being already today, and are going to be, light years ahead compared to the late 1970s, a common underlying conception links 1G with FN.

1.2 An Overview of Microtechnologies and Nanotechnologies (MEMS/NEMS)

After the high-level discussion in the previous pages, this section brings a sort of conceptual freefall, as the focus is going to be on HW technologies for low-complexity physical components, like sensors, actuators and transducers. Such a stretch is necessary to complete the deployment of key-elements, necessary for the reformulation of the concept of HW, which will be provided in the following chapters.

The set of HW solutions here at stake is the family of microtechnologies and nanotechnologies, intended as micro-/nano-devices, systems, also known by the MEMS and NEMS acronyms, along with innovative materials and nanomaterials. These items are envisaged as bearers of relevant potential to support, empower and enable functions and functionalities of 6G and FN, with particular emphasis on the edge of the network, the latter being the main purpose and proposition of this work.

It must be stressed that micro/nanotechnologies have been well-established in the scientific landscape for a long time. New research streams are always popping up, mainly driven by emerging applications needs, but MEMS/NEMS sensors, actuators and transducers, have been commercial products for about three decades. This said, the aim of this brief overview is not that of providing a detailed base of knowledge on the subject. The motivation for it is as follows. On one side, numerous remarkable articles and books are already available, and can easily be referred to for any in-depth analysis and such an effort would fall out of the scope of this book. In fact, the discussion to be developed around micro/nanotechnologies here is subordinated and functional to the redefinition of the concept of HW. Thereafter, the material reported in the continuation of this section will focus around the origin of micro/nanotechnologies, along with a few fundamental fabrication sequences, leaving a detailed review of the state-of-the-art uncovered, at least at this stage. In fact, the following chapters will feature, here and there, focused insights into existing works in literature, functional to support the main statements borne by this work.

1.2.1 Micro/nanotechnologies vs. semiconductors

The development of micro- and nanotechnologies has been following a path closely woven together with that of semiconductor technologies, yet marked by a few fundamental differences. Starting from the functional level, semiconductors, i.e., transistors, are active devices in the sense that, among other functions, they are capable of amplifying signals. Also, their characteristics of reconfigurability, like commuting from one state to another, tuning or sensing a certain magnitude, and so on, depend exclusively upon electrical/electronic properties, which are intrinsic to the (semiconductor) material. In other words, nothing is physically moving inside a semiconductor device.

On the other hand, MEMS and NEMS are passive components, as they only attenuate electrical signals. Plus, their sensing, actuating or transducing functionalities, always involve the physical deformation or displacement of part/parts of the device, e.g., a membrane or a suspended mass.

From the perspective of manufacturing, the fabrication process sequences of semiconductor and MEMS/NEMS devices have most part of the steps in common, as both rely on the selective deposition/removal of thin-films of metals, conducting and insulating materials. However, a limited set of processing steps differentiate them, as they are needed only by semiconductors or by MEMS/NEMS. These aspects will be covered in deeper detail in the following subsection.

Eventually, from the conceptual point of view, both semiconductors and MEMS/NEMS technologies pursue the trend to miniaturization. However, if CMOS transistors today are characterized by a channel length in the order of a few nanometers (despite the intrinsic device planar size being typically a few tens of nanometers), micro- and nano-device planar size can range from hundreds of nanometers to tens/hundreds of micrometers and up to a few millimeters. However, such a discrepancy needs to be better motivated. Semiconductors, and transistors in particular, have been following a seamless trend to miniaturization (i.e., scaling down), well-described by Moore's law [38]–[40]. In other words, with the advancement of technology and equipment for the fabrication of semiconductors, the same device, which is the transistor, was made smaller and smaller, with consequent advantages in terms of integrability, reduced power consumption, increasing speed, enhanced performance, etc. All this was possible thanks to high standardization of the manufacturing processes, with the employment of cutting edge and very expensive technologies, yet sustainable by reason of the large volumes of production. On the other hand, the development of micro/nanotechnologies followed a non-standardized path, which falls into the broad domain of more than Moore technologies [41]. This means that the rationale is not that of manufacturing smaller and smaller items, but to fabricate devices implementing diverse functions, and to obtain micro-devices (sensors, actuators, etc.) whose counterparts in traditional technologies can be as bulky and heavy as a brick. To provide an idea of how scarcely standardized are micro- and nanotechnologies with respect to unprecedented functions to be implemented, the saying "*One device, one technology*" [42], [43] became quite popular in the scientific community years ago when referring to MEMS and NEMS. For the matter of completeness, it must be highlighted that micro- and nano-devices, falling out of the Moore's law domain, can be successfully fabricated with equipment and technologies considered obsolete for decades in the CMOS industry. Eventually, semiconductors themselves can follow the more than Moore rationale. This is what happens, for instance, to heterostructures-based transistors [44–46], i.e., devices hybridizing silicon with other materials, like Ga, Ge, In, etc. In this context, scaling down is backed-up, favoring the exploration of performance and characteristics that cannot be achieved with standard (Moore's law compliant) solutions.

1.2.2 The early days of micro/nanotechnologies

As stated above, MEMS/NEMS and semiconductor technologies evolved having many things in common, while differentiating from each other for additional aspects.

From the point of view of technology, the pivotal processing steps of micro- and nano-devices developed together with the growth of semiconductors, starting in the 1950s. However, the utilization of those techniques to manufacture MEMS/NEMS commenced later, in the early 1970s. The advancement of silicon-based technologies pushed the investigation, besides electrical and electronic characteristics, toward the mechanical properties of the materials used in semiconductors manufacturing. In this sense, a few works, among others, deserving to be mentioned, analyze the mechanical properties of both bulk materials [47] and deposited thin layers [48–50], and were released between the mid-1950s and the mid-1960s. However, the utilization of these techniques for manufacturing micro-devices with movable parts and suspended thin membranes emerged later, from the second half of the 1970s to the early 1980s.

Mentioning a few more technical items, examples of anisotropic etching exploitation to yield various 3D suspended structures from a silicon substrate are discussed in [51]. This processing step, along with others usually adopted in the manufacturing of IC and transistors, brought to the first examples of miniaturized pressure sensors [52], accelerometers [53], [54], switches [55], [56], and other transducers for a variety of practical applications, touching, among others, the optical and biomedical fields. To this purpose, a relevant piece of work summarizing the state-of-the-art microsystem technologies, and providing a comprehensive outlook around diverse applications, was authored by Petersen [57] in the early 1980s.

Nevertheless, the realization of miniaturized systems in micro/nanotechnologies marked a change in pace later, thanks to the maturation of the surface micromachining fabrication technique [58]. This led to the modern conception of MEMS-based sensors and actuators, as we still intend them today. Given such a direction, a relevant contribution was authored by Howe and Muller [59] in 1983. The work reports on micro-cantilevers and double supported beams, realized in polycrystalline silicon, and suspended (after being released) above the substrate, exploiting silicon oxide as sacrificial layer. Since then, a wide variety of MEMS-based sensors, actuators, and various mechanisms, like gears and micro-motors, have been developed, tested, and reported in the scientific literature [60–62].

1.2.3 Fundamentals of micro-fabrication technologies

By virtue of what has been reported in the previous pages, it is trivial that the scientific area of micro- and nano-fabrication techniques and process technologies evolved, and is still evolving now, triggering a complex field of knowledge placed across physics, chemistry, mechanics, electronics, and so on. As already stated, the attempt of framing it into detail, would be scarcely relevant to the scope of this work, without even entering the amount of effort required. To this end, it is much effective referencing relevant contributions already available in the literature, such as [63–67].

Bearing in mind the targets of this book, a couple of relevant examples are going to be reviewed, with reference to two target MEMS devices manufactured by means of likewise technologies. The two processing flows at stake, namely bulk and surface micromachining, can be regarded as the pillars of micro-fabrication technologies, as process variations and evolutions always feature at least a few of their typical steps.

1.2.3.1 Bulk micromachining

In simple terms, a bulk micromachining process can be defined as a micro-fabrication sequence in which the substrate material (typically silicon) itself realizes the mechanical structure of the MEMS devices. In other words, the microsystem at stake, e.g., a sensor, is *made of silicon*.

The chosen example here is a MEMS-based EH device (EH-MEMS), which aims at converting part of the mechanical energy of vibrations into electricity by means of the piezoelectric transduction effect. The design concept is named FLC, after its resemblance to the clover (see Figure 1.8 for reference), and is extensively discussed in [68], [69]. The main steps of the bulk micromachining process employed for the fabrication of FLC EH-MEMS are wrapped together in Figure 1.7.

The substrates employed for fabrication are SOI wafers [70], i.e., two silicon layers, namely, the HLa (thicker) and the DLa (thinner), interfaced with a thin insulating layer (BOx). The first fabrication phase, reported in the sketches from

Figure 1.7: (a) SOI wafer. (b) Blank deposition of oxide. (c) Patterning of metal. (d) Passivation with oxide. (e) Oxide selective removal. (f) Piezoelectric layer patterning. (g) Patterning of metal. (h) Backside silicon etching. (i) Oxide removal. (j) Backside/frontside silicon etching. (k) Legend.

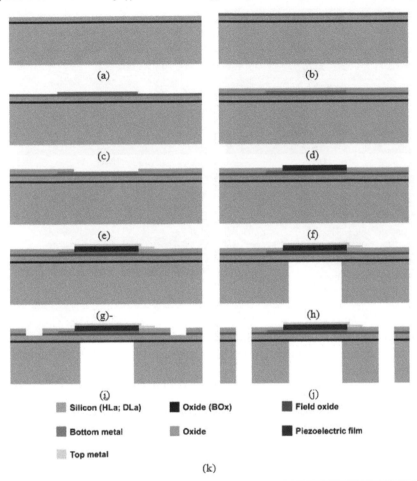

Figure 1.7 (a) to (g), consists of the selective deposition of two conductive layers, which are sandwiching a film of piezoelectric material (aluminum nitride). The former metals act as top and bottom electrodes, collecting the charges released by the piezoelectric transducers, when subjected to mechanical deformations. Thereafter, the second part of the processing, from Figure 1.7 (h) to (j), is the actual bulk micromachining, in which the substrate is selectively removed both on the front and back side of the wafer, thus shaping the silicon flexible suspensions and the proof mass, which realize the mass-spring mechanical

Figure 1.8: SEM images of the FLC EH-MEMS: (a) front side and (b) bottom side. Photographs of the FLC EH-MEMS: (c) bottom side and (d) front side of a couple of design variations mounted on a PCB.

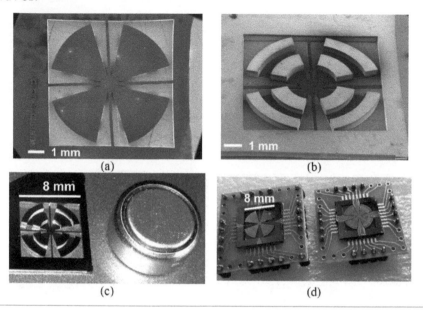

(a)   (b)

(c)   (d)

resonator, able to oscillate when subjected to mechanical vibrations. Front and back SEM images and a photograph of a physical FLC EH-MEMS sample are reported in Figure 1.8.

1.2.3.2 Surface micromachining

On the other side, a surface micromachining process can be defined as a micro-fabrication sequence in which the MEMS device is yield by a sequence of selective deposition and/or removal of thin-films, performed on the substrate material (typically silicon). In other words, the microsystem at stake, e.g., a sensors, is *made above silicon*.

The selected example is that of miniaturized RF passive components, like, e.g., switches, variable capacitors, phase shifters, variable attenuators, etc., realized in MEMS technology, broadly known as RF-MEMS [71], [72], [3]. The peculiar surface micromachining RF-MEMS technology at stake is detailed in [73–75], and yields electrostatically controlled devices. It is schematically shown in Figure 1.9.

Figure 1.9: a) Blank oxide deposition over silicon. (b) Patterning of polycrystalline silicon. (c) Oxide passivation. (d) Aluminum patterning. (e) Oxide passivation. (f) Opening of vias. (g) Evaporation of gold. (h) Sacrificial layer patterning. (i) Electrodeposition of gold. (j) Sacrificial layer removal. (k) Legend.

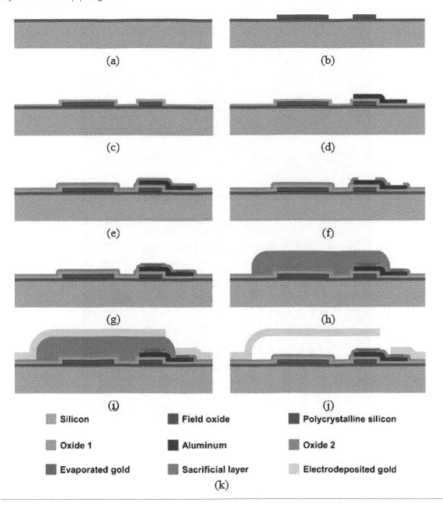

The fabrication starts with the deposition of a blank oxide layer over a silicon substrate (in Figure 1.9 (a)). Then, a polycrystalline silicon layer is selectively patterned, as shown in Figure 1.9 (b). Such a layer is used for DC biasing lines and electrodes and it is then passivated with oxide (Figure 1.9 (c)), before selective deposition of an aluminum layer (Figure 1.9 (d)), necessary for RF signal underpasses. Also the latter layer is protected by oxide

Figure 1.10: (a) Close-up (photograph) of a complex RF-MEMS device highlighting a few radially placed micro-switches. (b) Wider view of the device featuring RF-MEMS switches.

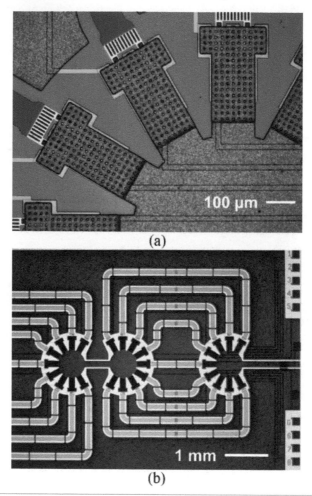

(a)

(b)

(Figure 1.9 (e)). Vias are opened in such an insulating layer, wherever electrical continuity has to be established with the underneath aluminum (Figure 1.9 (f)). Vias are then filled with evaporated gold (Figure 1.9 (g)). Thereafter, a sacrificial layer is patterned where suspended membranes are necessary (Figure 1.9 (h)). In fact, electrodeposition of gold structurally defines MEMS devices (Figure 1.9 (i)), whose movable parts are then released by removing the sacrificial layer (Figure 1.9 (j)). Microphotographs of fabricated RF-MEMS devices are reported in Figure 1.10.

1.3 Summary

This introductive chapter provided a set of preparatory concepts and information useful for grasping the main statements of this work, included in the following two chapters. First, a simplified, yet comprehensive description of the wireless communication standards, from 1G, to current 5G and future 6G and FN, was provided. This was done through the introduction of a graphical-based descriptive approach, called the CvE model. Then, the broad field of microtechnologies and nanotechnologies (MEMS/NEMS) was covered, focusing on their evolution, closely linked to that of standard semiconductor technologies. Eventually, the description of two pillar manufacturing approaches to the realization of micro- and nano-devices, namely, surface and bulk micromachining, was also provided.

References

[1] D. Tse and V. Pramod, *Fundamentals of wireless communication*. 2005

[2] W. C. Y. Lee, *Mobile communications design fundamentals*, 2nd ed. 2010

[3] J. Iannacci, *RF-MEMS Technology for High-Performance Passives – 5G applications and prospects for 6G*, 2nd ed. 2022

[4] A. Gupta, S. Member, R. K. Jha, and S. Member, "A survey of 5G network: Architecture and emerging technologies," *IEEE access*, vol. 3, 2015, doi: 10.1109/ACCESS.2015.2461602

[5] W. Saad, M. Bennis, and M. Chen, "A Vision of 6G Wireless Systems: Applications, Trends, Technologies, and Open Research Problems," *IEEE Netw.*, vol. 34, no. 3, 2020, doi: 10.1109/MNET.001.1900287

[6] *ITU IMT Vision—Framework and Overall Objectives of the Future Development of IMT for 2020 and Beyond*. Accessed: May. 18, 2023. [Online]. Available: https://www.itu.int/dms_pubrec/itu-r/rec/m/R-REC-M.2083-0-201509-I!!PDF-E.pdf

[7] *ITU Minimum Requirements Related to Technical Performancefor IMT-2020 Radio Interface(s)*. Accessed: May. 18, 2023. [Online]. Available: https://www.itu.int/dms_pub/itu-r/opb/rep/R-REP-M.2410-2017-PDF-E.pdf

[8] *The path to 5G: as much evolution as revolution*. Accessed: May. 18, 2023. [Online]. Available: https://www.3gpp.org/news-events/3gpp-news/5g-wiseharbour

[9] X. Lin, "An Overview of 5G Advanced Evolution in 3GPP Release 18," *IEEE Communications Standards Magazine*, vol. 6, no. 3. 2022, doi: 10.1109/MCOMSTD.0001.2200001

[10]   *Evolving LTE to fit the 5G future.* Accessed: May. 18, 2023. [Online]. Available: https://www.ericsson.com/en/reports-and-papers/ericsson-technology-review/articles/evolving-lte-to-fit-the-5g-future

[11]   *How the cellular ecosystem is evolving with the 4G to 5G transition.* Accessed: May. 18, 2023. [Online]. Available: https://www.ericsson.com/en/blog/2022/6/cellular-ecosystem-evolution-4g-to-5g-transition

[12]   *White Paper: 5G Network Architecture – A High-Level Perspective.* Accessed: May. 18, 2023. [Online]. Available: https://www-file.huawei.com/-/media/corporate/pdf/mbb/5g_nework_architecture_whitepaper_en.pdf?la=en

[13]   J. Kaur and M. A. Khan, "Sixth Generation (6G) Wireless Technology: An Overview, Vision, Challenges and Use Cases," 2022, doi:10.1109/TENSYMP545 29.2022.9864388

[14]   L. Zhang, Y. C. Liang, and D. Niyato, "6G Visions: Mobile ultra-broadband, super internet-of-things, and artificial intelligence," *China Commun.*, vol. 16, no. 8, 2019, doi:10.23919/JCC.2019.08.001

[15]   N. Khiadani, "Vision, Requirements and Challenges of Sixth Generation (6G) Networks," 2020, doi:10.1109/ICSPIS51611.2020.9349580

[16]   M. Z. Chowdhury, M. Shahjalal, S. Ahmed, and Y. M. Jang, "6G Wireless Communication Systems: Applications, Requirements, Technologies, Challenges, and Research Directions," *IEEE Open J. Commun. Soc.*, vol. 1, 2020, doi:10.1109/ojcoms.2020.3010270

[17]   A. Mourad, R. Yang, P. H. Lehne, and A. De La Oliva, "Towards 6G: Evolution of key performance indicators and technology trends," 2020, doi: 10.1109/6GSUMMIT49458.2020.9083759

[18]   Z. L. Yi, S. Wang, S. F. Han, C. F. Cui, and Y. F. Wang, "From 5G to 6G: Requirements, Challenges and Technical Trends," *Beijing Youdian Daxue Xuebao/Journal of Beijing University of Posts and Telecommunications*, vol. 43, no. 2. 2020, doi: 10.13190/j.jbupt.2020-024

[19]   H. Wymeersch *et al.*, "6G Radio Requirements to Support Integrated Communication, Localization, and Sensing," 2022, doi: 10.1109/EuCNC/6GSummit54941.2022.9815783

[20]   S. A. Abdel Hakeem, H. H. Hussein, and H. W. Kim, "Vision and research directions of 6G technologies and applications," *Journal of King Saud University - Computer and Information Sciences*, vol. 34, no. 6. 2022, doi: 10.1016/j.jksuci.2022.03.019

[21]   X. An, J. Wu, W. Tong, P. Zhu, and Y. Chen, "6G network architecture vision," 2021, doi: 10.1109/EuCNC/6GSummit51104.2021.9482439

[22]   Z. Zhang *et al.*, "6G Wireless Networks: Vision, Requirements, Architecture, and Key Technologies," *IEEE Veh. Technol. Mag.*, vol. 14, no. 3, 2019, doi: 10.1109/MVT.2019.2921208

[23] L. Bariah *et al.*, "A prospective look: Key enabling technologies, applications and open research topics in 6G networks," *IEEE Access*, vol. 8, 2020, doi: 10.1109/ACCESS.2020.3019590

[24] B. Zong, C. Fan, X. Wang, X. Duan, B. Wang, and J. Wang, "6G Technologies: Key Drivers, Core Requirements, System Architectures, and Enabling Technologies," *IEEE Veh. Technol. Mag.*, vol. 14, no. 3, 2019, doi: 10.1109/MVT.2019.2921398

[25] C. De Alwis *et al.*, "Survey on 6G Frontiers: Trends, Applications, Requirements, Technologies and Future Research," *IEEE Open J. Commun. Soc.*, vol. 2, 2021, doi: 10.1109/OJCOMS.2021.3071496

[26] X. You *et al.*, "Towards 6G wireless communication networks: vision, enabling technologies, and new paradigm shifts," *Science China Information Sciences*, vol. 64, no. 1. 2021, doi: 10.1007/s11432-020-2955-6

[27] J. F. Valenzuela-Valdés, Á. Palomares, J. C. González-Macías, A. Valenzuela-Valdés, P. Padilla, and F. Luna-Valero, "On the Ultra-Dense Small Cell Deployment for 5G Networks," 2018, doi: 10.1109/5GWF.2018.8516948

[28] M. Latva-Aho, A. Pouttu, A. Hekkala, I. Harjula, and J. Mäkelä, "Small cell based 5G test network (5GTN)," in *Proceedings of the International Symposium on Wireless Communication Systems*, 2015, vol. 2016-April, doi: 10.1109/ISWCS.2015.7454335

[29] R. K. Saha, "Approaches to Improve Millimeter-Wave Spectrum Utilization Using Indoor Small Cells in Multi-Operator Environments Toward 6G," *IEEE Access*, vol. 8, 2020, doi: 10.1109/ACCESS.2020.3037684

[30] R. K. Saha, "Underlay cognitive radio millimeter-wave spectrum access for in-building dense small cells in multi-operator environments toward 6G," in *International Symposium on Wireless Personal Multimedia Communications, WPMC*, 2020, vol. 2020-October, doi: 10.1109/WPMC50192.2020.9309471

[31] A. Mukherjee, D. De, N. Dey, R. G. Crespo, and E. Herrera-Viedma, "DisastDrone: A Disaster Aware Consumer Internet of Drone Things System in Ultra-Low Latent 6G Network," *IEEE Trans. Consum. Electron.*, vol. 69, no. 1, 2023, doi: 10.1109/TCE.2022.3214568

[32] D. Mishra, A. M. Vegni, V. Loscri, and E. Natalizio, "Drone Networking in the 6G Era: A Technology Overview," *IEEE Commun. Stand. Mag.*, vol. 5, no. 4, 2021, doi: 10.1109/MCOMSTD.0001.2100016

[33] R. C. Moioli *et al.*, "Neurosciences and wireless networks: The potential of brain-type communications and their applications," *IEEE Commun. Surv. Tutorials*, vol. 23, no. 3, 2021, doi: 10.1109/COMST.2021.3090778

[34] J. Mertes *et al.*, "Evaluation of 5G-capable framework for highly mobile, scalable human-machine interfaces in cyber-physical production systems," *J. Manuf. Syst.*, vol. 64, 2022, doi: 10.1016/j.jmsy.2022.08.009

[35] M. -K. Kim, J. -H. Cho, H. -B. Shin and S. -W. Lee, "Towards Brain-based Interface for Communication and Control by Skin Touch," in *11th International Winter Conference on Brain-Computer Interface (BCI)*, 2023, doi: 10.1109/BCI57258.2023.10078458

[36] J. Iannacci, "A Perspective Vision of Micro/Nano Systems and Technologies as Enablers of 6G, Super-IoT, and Tactile Internet," *Proc. IEEE*, vol. 111, no. 1, 2023, doi: 10.1109/JPROC.2022.3223791

[37] J. Iannacci and H. V. Poor, "Review and Perspectives of Micro/Nano Technologies as Key-Enablers of 6G," *IEEE Access*, vol. 10, 2022, doi: 10.1109/ACCESS.2022.3176348

[38] P. K. Bondyopadhyay, "Moore's law governs the silicon revolution," *Proc. IEEE*, vol. 86, no. 1, 1998, doi: 10.1109/5.658761

[39] G. Strawn and C. Strawn, "Moore's Law at Fifty," *IT Prof.*, vol. 17, no. 6, 2015, doi: 10.1109/MITP.2015.109

[40] M. Golio, "Fifty Years of Moore's Law," *Proc. IEEE*, vol. 103, no. 10, 2015, doi: 10.1109/JPROC.2015.2473896

[41] A. B. Kahng, "Scaling: More than Moore's law," *IEEE Des. Test Comput.*, vol. 27, no. 3, 2010, doi: 10.1109/MDT.2010.71

[42] J. C. Eloy, E. Mounier, "Status of the MEMS industry," in Proc. of SPIE 5717, MEMS/MOEMS Components and Their Applications II, 2005, doi: 10.1117/12.594011

[43] M. F. Niekiel *et al.*, "What MEMS Research and Development Can Learn from a Production Environment," MDPI Sensors, vol. 23, no. 12, pp. 1–25, 2023, doi: 10.3390/s23125549

[44] C. Lamberti and G. Agostini, *Characterization of semiconductor heterostructures and nanostructures.* 2013

[45] H. F. Huang *et al.*, "Investigation of a GaN-on-Si HEMT optimized for the 5th-generation wireless communication," 2016, doi: 10.1109/ASICON.2015.7517068

[46] R. Yang, R. Xu, W. Dou, M. Benner, Q. Zhang, and J. Liu, "Semiconductor-based dynamic heterojunctions as an emerging strategy for high direct-current mechanical energy harvesting," *Nano Energy*, vol. 83. 2021, doi: 10.1016/j.nanoen.2021.105849

[47] J. H. Hobstetter, *Mechanical properties of semiconductors Properties of Crystalline Solids*, 1st ed. 1960

[48] J. W. Beams, J. B. Breazeale, and W. L. Bart, "Mechanical strength of thin films of metals," *Physical Review*, vol. 100, no. 6. pp. 1657–1661, 1955, doi: 10.1103/PhysRev.100.1657

[49] C. A. Neugebauer, "Tensile properties of thin, evaporated gold films," *J. Appl. Phys.*, vol. 31, no. 6, 1960, doi: 10.1063/1.1735751

[50] J. M. Blakely, "Mechanical properties of vacuum-deposited gold films," *J. Appl. Phys.*, vol. 35, no. 6, 1964, doi: 10.1063/1.1713735

[51]   E. Bassous, "Fabrication of novel three-dimensional microstructures by the anisotropic etching of (100) and (110) silicon," *IEEE Trans. Electron Devices*, vol. 25, no. 10, 1978, doi: 10.1109/T-ED.1978.19249

[52]   W. H. Ko, J. Hynecek, and S. F. Boettcher, "Development of a Miniature Pressure Transducer for Biomedical Applications," *IEEE Trans. Electron Devices*, vol. 26, no. 12, 1979, doi: 10.1109/T-ED.1979.19793

[53]   L. M. Roylance and J. B. Angell, "A Batch-Fabricated Silicon Accelerometer," *IEEE Trans. Electron Devices*, vol. 26, no. 12, 1979, doi: 10.1109/T-ED.1979.19795

[54]   P. Chen, R. S. Muller, T. Shiosaki, and R. M. White, "WP-B6 Silicon Cantilever Beam Accelerometer Utilizing a PI-FET Capacitive Transducer," *IEEE Trans. Electron Devices*, vol. 26, no. 11, 1979, doi: 10.1109/T-ED.1979.19782

[55]   K. E. Petersen, "Micromechanical membrane switches on silicon," *IBM J. Res. Dev.*, vol. 23, no. 4, 1979, doi: 10.1147/rd.234.0376

[56]   K. E. Petersen, "Micromechanical light modulator array fabricated on silicon," *Appl. Phys. Lett.*, vol. 31, no. 8, 1977, doi: 10.1063/1.89761

[57]   K. E. Petersen, "Silicon as a mechanical material," *Proc. IEEE*, vol. 70, no. 5, pp. 420–457, 1982, doi: 10.1109/PROC.1982.12331

[58]   J. M. Bustillo, R. T. Howe, R. S. Muller, and L. Fellow, "Surface Micromachining for Microelectromechanical Systems," vol. 86, no. 8, 1998

[59]   R. T. Howe and R. S. Mueller, "Polycrystalline Silicon Micromechanical Beams," *J. Electrochem. Soc.*, vol. 130, no. 6, p. 1420, 1983, doi: 10.1149/1.2119965

[60]   R. T. Howe and R. S. Muller, "Resonant-Microbridge Vapor Sensor," *IEEE Trans. Electron Devices*, vol. 33, no. 4, 1986, doi: 10.1109/T-ED.1986.22519

[61]   *Pin joints, gears, springs, cranks, and other novel micromechanical structures*. Accessed: May. 29, 2023. [Online]. Available: https://www.osti.gov/biblio/5974687

[62]   Y. C. Tai and R. S. Muller, "IC-processed electrostatic synchronous micromotors," *Sensors and Actuators*, vol. 20, no. 1–2, 1989, doi: 10.1016/0250-6874(89)87101-X

[63]   E. Kondoh, *Micro- and Nanofabrication for Beginners*. 2022

[64]   W. C. O'Mara, R. B. Herring, L. P. Hunt, *Handbook of semiconductor silicon technology*, 1990

[65]   W. M. Moreau, *Semiconductor Lithography: Principles, Practices, and Materials*. 1988

[66]   M. J. Madou, *Fundamentals of Microfabrication and Nanotechnology, Three-Volume Set*. 2018

[67]   H. H. Gatzen, V. Saile, and J. Leuthold, *Micro and nano fabrication: Tools and processes*. 2015

[68]   J. Iannacci *et al.*, "Multi-modal vibration based MEMS energy harvesters for ultra-low power wireless functional nodes," *Microsyst. Technol.*, no. 1998, Dec. 2013, doi: 10.1007/s00542-013-1998-2

[69]   J. Iannacci, G. Sordo, E. Serra, and U. Schmid, "The MEMS four-leaf clover wideband vibration energy harvesting device: design concept and experimental verification," *Microsyst. Technol.*, vol. 22, no. 7, 2016, doi: 10.1007/s00542-016-2886-3

[70]   M. Tilli, T. Motooka, V. M. Airaksinen, S. Franssila, M. Paulasto-Kröckel, and V. Lindroos, *Handbook of Silicon Based MEMS Materials and Technologies: Second Edition.* 2015

[71]   G. M. Rebeiz, *RF MEMS: Theory, Design and Technology.* 2003

[72]   J. Iannacci, *Practical Guide to RF-MEMS.* 2013

[73]   F. Giacomozzi, V. Mulloni, S. Colpo, J. Iannacci, B. Margesin, and A. Faes, "A flexible fabrication process for RF MEMS devices," *Rom. J. Inf. Sci. Technol.*, vol. 14, no. 3, 2011

[74]   J. Iannacci and C. Tschoban, "RF-MEMS for future mobile applications: Experimental verification of a reconfigurable 8-bit power attenuator up to 110 GHz," *J. Micromechanics Microengineering*, vol. 27, no. 4, 2017, doi: 10.1088/1361-6439/aa5f2c

[75]   J. Iannacci, "RF-MEMS technology as an enabler of 5G: Low-loss ohmic switch tested up to 110 GHz," *Sensors Actuators, A Phys.*, vol. 279, 2018, doi: 10.1016/j.sna.2018.07.005

2

An Abstract Playground to Reformulate the Concept of Hardware in View of 6G and FN

Abstract

This chapter makes an effort to motivate the need for a reformulation of the concept of HW, in view of 6G and FN, and of their functioning. This undertaking will build upon the development of a conceptual framework, which is also provided with a graphical 3D implementation. The ad-hoc-constructed tool is named the I2D model, and is based on a deductive approach. The I2D model features three sets of elements, which can also be visualized as dimensions laying in a three-dimensional space. The first is that of CMR, which are high-level needs and KPI, increasingly important for the transition to 6G/FN. The second dimension is that of TMT. These are technology trends that will be driven by CMR, i.e., when moving from high-level KPI towards the actual physical (HW–SW) implementation and deployment of the network. Finally, the third (vertical) dimension of the I2D model is that of 3F. Taking a further step down, i.e., towards the physical realization of components, 3F are actual functionalities that, putting together the demands outlined by CMR and TMT, future low-complexity HW devices will be expected to satisfy. The third dimension of 3F closes the loop with microtechnologies and nanotechnologies. As the last item, the chapter will report a set of correlation matrices, unravelling the set of interdependences existing among CMR, TMT and 3F.

2.1 The Identifying Defining Deriving (I2D) Methodology

The previous chapter sketched scenarios of relevant complexity. The paradigms of 6G and FN highlight, themselves, sets of ramifications at various levels, e.g., technologies, applications, services and protocols, which are practically impossible to mentally embrace at once. In addition, the discussion of micro/nanotechnologies, placed significantly *below* in terms of complexity, added further elements that need to be harmonized. The scope of this chapter is to bring together all the just mentioned items, building the core proposition conveyed by the book. In other terms, *why* and *how* the standard concept of HW needs to be revised and updated to fully meet the expectations of 6G/FN, are going to be addressed here, also highlighting the pivotal role of micro- and nanotechnologies in this unprecedented (and still unknown to a certain extent) transition.

The way this objective is pursued is via the definition of a methodology that helps infer the practical needs, in terms of KPI, which will have to be implemented, deriving them according to a top-down approach, i.e., from system-level requirement, to component-level specifications. Once this task is accomplished, KET, both existing and to be developed/enhanced, capable of meeting the challenge, will be more easily identifiable.

In order to keep the understanding of the methodology as simple as possible, a 3D conceptual space is introduced. It is named I2D, which stands for identifying defining deriving. Such a three-step model starts with the identifying (I) phase, in which the emerging high-level KPI of 6G and FN are grouped in a set of trends, named critical mega-requirements (CMR). Once this step is fixed, the subsequent defining (D) phase aims to *translate* CMR, which are high-level entities, into technology trends, which are middle-level items. The outcome of such a second phase is a set of technology trends named technology mega-trends (TMT). Notably, TMT are predominantly linked to HW technologies, providing a first glimpse into the relevance of bottom-up approaches, i.e., from the device-, to the system-, to the service- and application-level, in the emerging 6G/FN contexts. Eventually, the third and last deriving (D) phase puts together CMR and TMT, leading to a set of actual functionalities that low-complexity HW components are expected to implement in the future, with particular emphasis on the network edge. Such items are named forecasted functional features (3F). The introduced items are collectively put together in the 3D conceptual space shown in Figure 2.1.

Conceptually, a 2D ground plane is sketched by 6G/FN CMR, i.e., application-/service-level key demands, and TMT, i.e., HW/SW technologies and

Figure 2.1: Conceptual 3D matrix matching CMR, TMT and 3F, as output of the I2D methodology. (All the images and thumbnails used to compose the graphic are powered by Freepik.com.)

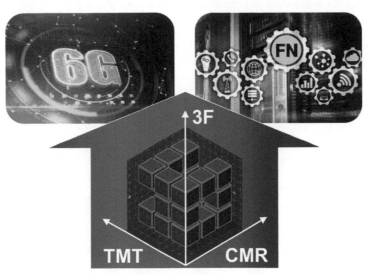

solutions, which can address one or more CMR. The vertical dimension is that of 3F, which are desired operation characteristics to be implemented by next generation HW simple devices, like, e.g., sensors and actuators. 3F bring CMR and TMT together, and build proper ground for 6G and FN to further develop.

2.2 Identifying (I) Critical Mega-requirements (CMR)

As already discussed, the first identifying (I) phase of the I2D model leads to the formulation of high-level requirements of 6G and FN, which can also be appropriately interpreted as service- and functional-level KPI. Such items are here named CMR, and are listed, one-by-one, in the following subsections.

2.2.1 Volume of data over the network (CMR-1)

Recalling the scenario sketched in Chapter 1, data-related KPI are constantly increasing, following the significant climb from 4G, to break the 1 Gbps frontier of average data rate per user under 5G [1] and to improve such a number by 100–1000 times (i.e., Tbps) with the uptake of 6G [2], [3], in around 2030. Beside the improvement of QoS and the pervasivity of the functions offered to the end user (service plane), pivotal technologies that are being and will be employed for the network functioning (operation plane), like NFV [4] and AI [5], are also demanding for more data, more reliability and less latency.

The resulting scenario is that of a relentless increase of data traffic over the network, with forecasted numbers confirmed afterwards by actual data. These considerations are supported by the data in Figure 2.2.

Figure 2.2: (a) Mobile data traffic in Western Europe forecast from 2016 to 2021 [6]. (b) Quarterly data traffic in Spain from 2013 to 2021 [7].

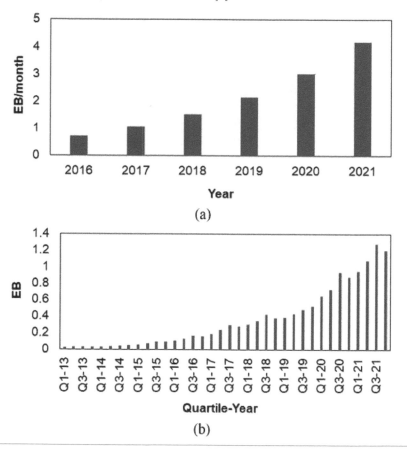

(a)

(b)

The histogram in Figure 2.2 (a) refers to the overall volume of data traffic in Western Europe (in EB/month), forecasted in 2016 for the next years to 2021 [6]. This means that the number for 2016 is based on actual data, while the others are predictions. The reported information include file sharing, video streaming, plus data generated by M2M applications, laptops, tablets and other portable devices (including wearables). The emerging trend is undeniable, despite being based on forecasts. To this end, it is interesting to observe how such growth is confirmed by the factual (not predicted) data in Figure 2.2 (b). This is the quarterly data traffic in mobile broadband networks in Spain (in EB) from 2013 to 2021 [7]. Confronting the data in the two graphs, the predicted growth of about a factor of five between 2016 and 2021 in Western Europe is confirmed by an increase of around six times, across the same timeframe, in the whole of Spain. Along this direction, the total amount of data created worldwide is forecasted to climb from 64 ZB in 2020 to around 170 ZB in 2025 [8].

2.2.2 Centralization vs. distribution of intelligence (CMR-2)

The previous CMR-1 on data traffic increase is a mega-requirement pushed by any advancement of the network, as well as of its services and functionalities, regardless of whether they are located more in the vicinity of the core or of the edge. By contrast, the CMR introduced here is closely linked to the trends to a relentless increase of functions distributed to the edge of the network, already mentioned in Chapter 1. In addition, the proliferation and capillarization of services is going to be radical to such an extent that all the following CMR are driven by it, with respect to diverse levels, like data storage, elaboration, energy, and so forth.

Mentioning numbers, the density of devices connected to the network per square kilometer of 0.1 M/km^2 under 4G, is expected to rise to 1 M/km^2 with 5G, and to 10 M/km^2 when 6G will take over [9], as shown in Figure 2.3.

Figure 2.3: Millions of devices per square kilometer from 4G to 6G [9].

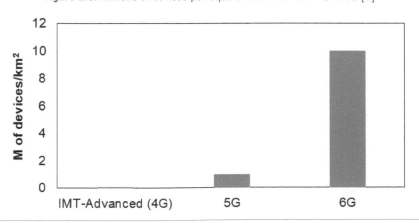

Bearing in mind such a context, the number of IoT devices connected to the network, spanning from home, to smart-city, to industrial and more applications, is expected to break 24 billion in 2025 [8]. In addition, M2M communications will also contribute to increasing the overall amount of data traveling across the network.

Recalling the 170 ZB of data estimated in 2025 [8], which is by itself a massive amount of information traveling across the network, handling IoT- and M2M-generated data with a traditional centralized approach would lead to even larger numbers, and, conversely, to a scenario scarcely manageable and sustainable.

Putting together such conflicting trends, it is straightforward that concurrent centralization and distribution of intelligence, from the core to the edge, are the two leverages to act upon, to make services more efficient, faster, ubiquitous, readily accessible and reliable, while limiting their weight on data traffic volumes [10].

2.2.3 Edge intelligence (EI) and local self-management (CMR-3)

As has already emerged, pursuing the trend to distribution brings a set of important implications. Starting from a rather high level of abstraction, implementing functions and services at the tinier ramifications of the network means bringing *intelligence* to the edge. In this sense, the term intelligence must be interpreted quite broadly, comprehending, among other things, computational capacities, storage of information, connectivity to the rest of the network, and so on. Bearing this in mind, today it is very common discussing about EI and edge computing, which are, in a way, equivalent terms. Edge computing is the key to enable self-management features of small portions of the network, while maintaining the ability to work as a whole with the rest of the infrastructure [11–13]. Also, EI is the cornerstone of the so-called "Third act of the internet" [14], [15]. In a nutshell, looking at the evolution of the internet, from the 1990s to date, the first act was related to connecting computers together, everywhere in the world. The second act, still ongoing today, is hinged around remotely accessing content and data that are centrally stored. This is the case, for instance, of video streaming, gaming, social media, and so on. The third act, instead, has to do with EI and distribution of services to the edge.

One common reference example to easily understand the difference between the current centralization of services against what EI could do in the future is that of smart speakers. These edge-items, which we got used to in our homes, perform very little or no elaboration of data locally. What they do, instead, is to record our voice, and send such information somewhere toward the core of the network, where intelligence and data storage capacities are located. When the voice is analyzed and the action to be taken in response is identified, this instruction is transmitted back, from the core to the smart speaker. What the item will then do most probably requires once again broadcasting data back and forth across the network. Diversely, a smart speaker based on EI would implement locally most part of the elaboration, minimizing the effect upon the generated data traffic, yet maintaining a proper link to the rest of the infrastructure, to work as a whole rather than as an isolated system.

Eventually, just to put together some clues around the relevance of EI/edge computing, the forecasted market volume of applications related to such topics is expected to reach worldwide USD274 billion by 2025 [16].

2.2.4 Sensing and conditioning at the edge (CMR-4)

It is straightforward that implementing EI, which is a rather high-level functional concept, demands a variety of lower-complexity (physical) components and devices. Thereafter, starting from the current CMR, the attention is going to be oriented towards the identification of classes of HW components, necessary to empower the realization of the EI paradigm.

The CMR highlighted here has to do with sensing and conditioning functionalities. In fact, devices, regardless of their ability to perform edge computing, have first to detect local and environmental magnitudes, before acting or reacting in any way.

The closest example we have at hand is the smartphone. In fact, it is provided with a variety of sensors, actuators and transducers, like inertial, temperature, light, pressure sensors, along with fingerprint, biometric sensors, and other devices [17]. If we consider for a moment the smartphone as an edge device, it is equipped with relevant edge computing capacities. To this end, it is more appropriate categorizing as edge items devices like wearables (e.g., smartwatches, body sensors, smart glasses, etc.) and remote sensing nodes (in homes, cities, factories, etc.).

In any case, recalling the expected numbers of devices per square kilometer in Figure 2.3, and considering that each of those devices will have to be equipped with at least a few, among sensors, actuators and transducers, it is easy to figure out the volume of such physical items that will be necessary in the future.

2.2.5 Data storage and computation at the edge (CMR-5)

Another crucial class of physical items needed to enable EI is that of data storage and computation components. Taking again the smartphone as a temporary reference, it can be referred to as a rough example of an edge device, capable of storing and elaborating data. However, the smartphone has an architecture similar to that of a computer, and its memory and computation performance are excessive for the scopes of the edge computing devices here at stake. In fact, smartphones, along with computers, are multi-purpose and general-purpose devices, able to support the (parallel) execution of different services, algorithms, software applications and apps.

More effectively, when dealing with edge devices, whose complexity can range from a single remote sensor, to systems for monitoring living spaces, metropolitan areas, factories and appliances, again, to wearable and implantables, memory and computation capacities must be tailored to the complexity of the function/s to be implemented. In other words, one should imagine EI implemented by a specific (rather than general) purpose physical system. This is the case, e.g., of ASICs, which are highly customized computation modules [18], [19].

Apart from the trivial advantages of having an amount of intelligence that is appropriately sized against the function/s to be implemented, e.g., in terms of power consumption, operation efficiency, stability/reliability, etc., one should always bear in mind the numbers in Figure 2.3.

2.2.6 Energy distribution for edge operation (CMR-6)

The more intuitive demands in terms of physical components required by a concurrent increase of distribution and proliferation of edge computing, were covered by previous CMR. Though, nested in such an unprecedented paradigm transition, there exists another CMR that probably is not so easy to figure out, as often it is taken for granted. Yet, it can swing the balance towards full or partial feasibility of ubiquitous EI. The CMR at stake here is that of energy availability for operation.

It is enough to observe what happens in our daily lives, to grasp the importance of energy for device operation. With the widespread deployment of 2G, characterized by very-efficient digital coding techniques, along with handsets just for calling and texting, with tiny low-power LCD screens, we got used to mobile devices needing to be plugged in for recharging probably once a week. Today, instead, we have smartphones equipped with power-hungry devices, like high-resolution screens, loudspeakers, etc., which are extensively exchanging data with the infrastructure (e.g., video streaming), demanding battery recharge every day and everywhere (cars, airports, shopping malls, and so on). Also relevant, we have gotten used to a plethora of other portable and connected devices, like smart watches, smart speakers, smart meters/sensors, e.g., wearable or placed on our car, bicycle, etc., which all come along with a battery charger or a cable for USB charging.

Keeping this in mind, it is easier to get the complexity of powering millions of smart devices per square kilometer (see Figure 2.3). On one hand, as emerged before, EI will be based on optimized and energy-efficient systems. Yet, from a different perspective, it is simply not an option to wire such devices with power cables, or have them relying on batteries that at some point need to be replaced. The scientific community is well aware of these issues, and research is already oriented to their handling [20–24].

2.2.7 Orchestration of a network of networks (CMR-7)

Wrapping together what was previously discussed, decentralization and scattering of EI will increase quite relevantly the autonomy and self-management of physical nodes placed at the tinier ramifications of the network infrastructure. In this sense, it is becoming common to address the B5G, 6G and FN scenario with the term *network of networks* [25]. As already highlighted, the network will evolve from the current centralized approach of services and data, to their massive decentralization and distribution at the edge. However, as stressed in the CvE description in Chapter 1, this does not mean that centralization will simply be dropped. In fact, a non-negligible level of it will have to be maintained and also expanded in order to make the whole network work as a whole. To better grasp the complexity behind it, one should focus on the difference between what we can define as traditional vs. evolved centralization. What has been increasingly pursued across the previous generations, is a sort of repositioning of *intelligence* (still used with the broadest possible meaning) from the edge to the core of the network. Reasoning in terms of pros and cons, this approach had undeniable advantages, like the reduction of HW redundancy, the increase of performance and functionalities, synchronization and consistency of data, ease of maintenance and of services upgrade, etc. On the other hand, this took place at the cost of a massively increased load on Tx/Rx data traffic volumes, increased latency, reliability issues depending on the available bandwidth and coverage, as well as unavailability of services wherever coverage is absent.

From a different perspective, the novel conception of centralization will have to manage edge nodes that are less burdensome in terms of Tx/Rx data, since an increased portion of computation and data storage is locally managed. Yet, by virtue of their resident intelligence, such nodes are more autonomous, and able to deploy self-management and self-evolving strategies. As such, it is straightforward that the challenges to be addressed by future centralization strategies are not trivial. It fact, the scientific research is already very active in developing novel approaches oriented to the network *orchestration* [26–34], so that local autonomy will not turn into an impairment to work as a whole, from its core to the edge.

2.3 Defining (D) Technology Mega-trends (TMT)

The comprehensive discussion developed in the previous section completed the identifying (I) phase of the I2D model, reviewing CMR and consolidating one of the three dimensions of the conceptual 3D matrix in Figure 2.1. Starting from such a base of knowledge, the second phase, named defining (D), is going to be addressed. As mentioned earlier, the objective of this analysis is to transform CMR, which are high-level and vertical entities, into horizontal technology trends. In other terms, the technology mega-trends (TMT) defined in the following, represent features and characteristics that are desirable for 6G and FN applications, regardless of the specific HW technology and device at stake.

2.3.1 Increase of miniaturization (TMT-1)

The common attitude in linking technology advancement with enabling the realization of smaller and smaller physical items, is one of the cornerstones of our time. Tracing back the origin of this is a rather simple task. For decades, we have been surrounded by electronic devices and appliances featuring electronic parts. The invention of the transistor, announced in 1948 by John Bardeen, Walter Brattain and William Shockley at Bell Labs [35], represented itself an abrupt step towards scaling down. In fact, tiny semiconductor-based devices, replaced the traditional bulky vacuum tubes. Then, a relentless trend to miniaturization, still ongoing today, after several decades, has been pursued by the semiconductor industry, as described by Moore's law [36] (see Chapter 1). A similar approach is also followed by non-silicon technologies, despite at a different pace, rules and constraints. This is, for example, the case of microtechnologies and nanotechnologies, which fall into the so-called more than Moore path [37] (see Chapter 1). Sensors, actuators and transducers in MEMS/NEMS technologies are unavoidably larger if compared to state-of-the-art transistors, but are able to realize more sophisticated and diversified functions and, in addition, are way smaller and less bulky than traditional counterparts. To this end, just to pick an example, one should focus on the difference in size between classical and MEMS-based accelerometers, gyroscopes and IMU [38]. That said, and also considering the very-close intricacies between electronics and telecommunications, further increase of miniaturization will be one of the main pillars of technologies for 6G and FN.

2.3.2 Increase of integration (TMT-2)

In rather close conjunction with miniaturization, more than Moore technologies triggered the trend to integration. In simple terms, when the need to realize more complex functions arises, it is necessary to put together an increasing number of components. To do so, one has two main options. On one hand, those items can be monolithically manufactured within the same technology platform, through a unique processing sequence (Moore's law compliant approach known as SoC) [39], [40]. Depending on the specific system or sub-system targeted, this approach can be pursued to a limited extent. In fact, reaching a certain complexity with monolithic chips can mean increasing the manufacturing complexity and cost, reducing the yield, volumes, increasing (on silicon) area consumption and time for production. Also, not necessarily a unique technology is the best solution to realize all the components to be incorporated into a system or sub-system.

The other option available is that of integrating components realized in different and incompatible technologies, within a unique system, realizing what is often named as heterogeneous integration (more than Moore approach). In practical terms, such components are mounted together on a board or a carrier substrate, thus realizing a more complex device [41]. To this purpose, packaging technologies, originally devoted mainly to protect and seal single components, are providing great boost in easing and enabling unprecedented integration solutions [42]. In fact, nowadays it is common to exploit the package itself as housing and carrier of components, facilitating their interfacing, and leading to the concept of SiP and 3D-stacks [43–46].

2.3.3 Increase of operation differentiation and diversity (TMT-3)

The TMT that is going to be covered here, along with the following TMT-4 and TMT-5, realizes convergence of different factors. In fact, if they are increasingly pushed by emerging application demands, on the other hand, they are inherent when pursuing miniaturization (TMT-1) along with integration (TMT-2). Focusing now on the diversity and differentiation of operation and functionalities, it is straightforward that modern electronic systems and sub-systems are able to implement different services and to work in different states (reconfigurability), relying on the same HW, thus reducing physical redundancy, cost, area occupation, and so on. This is possible thanks to the combination of two factors, i.e., the evolution of semiconductor technologies and, on the other hand, the virtualization of functions [47–49]. Such a scenario is currently turning into reality, under the auspices of 5G, at various levels of abstraction and complexity of the infrastructure.

The just-depicted scenario mainly refers to HW systems and sub-systems. The actual disruption here addressed is that the current TMT also comprises low-complexity physical components, like sensors and actuators. In more explicit terms, the EI capillarity of 6G and FN will demand that elemental HW devices are able to diversify their operation and functionality, e.g., a pressure sensor that can also function as an inertial sensor, as a micro-relay, and so on.

2.3.4 Increase of redundancy (TMT-4)

At first sight, this TMT seems to be in opposition to what was discussed to this point. At higher-level, evolution of services is moving towards a reduction of redundancy, thanks to broader exploitation of the available resources, as well as relying on open models [50–53]. In addition, something similar is taking place at the low-complexity level, i.e., for what concerns simple physical components. In fact, actual conflicts do not build, as the reduction of redundancy at the infrastruc-ture architectural level brings extensive simplification, which is not jeopardized, not even to the smaller extent, by local increase of HW redundancy at the edge. The latter statement can be further supported by highlighting that a certain level of HW redundancy at the edge is accompanied by very-limited impact at various levels, like, e.g., cost, area occupation, energy consumption, etc. This is made possible by the relentless increase of miniaturization (TMT-1) and integration (TMT-2) of technologies. On the other hand, the increase of HW redundancy at the edge is a fundamental key to enable differentiation and diversity of operation (TMT-3), along with other core targets of 6G and FN, like high-resilience, high-reliability and self-management of small segments of the network.

2.3.5 Decrease of power consumption (TMT-5)

This TMT, along with the subsequent two, addresses the topic of energy. In particular, the three TMT focus on energy consumption, procurement and availability. In the current discussion, the trend towards the decrease in power consumption is at stake. As already mentioned, TMT-5 falls into the set of trends that are both driven by emerging requirements, and intrinsically enabled by miniaturization and integration. That said, it is straightforward that recent generations of electronic circuits, blocks and components, are less and less power hungry if compared to similar designs in previous technologies, while marking, at the same time, increased performance. From a different perspective, proliferation of remote nodes for sensing, conditioning and other purposes, pushed by applications paradigms like the IoT, is posing demanding constraints in terms of power consumption [54–56], given the limited availability of local energy, along with the difficulty to reach with power cables or for battery replacement.

2.3.6 Harvesting of energy from the environment (TMT-6)

The decrease in power consumption (TMT-5) is undoubtedly a crucial factor, yet it is not enough to enable widespread distribution of intelligence and functionalities at the edge. In fact, limited power demands can extend the operability of remote nodes, but not relieve them from passing out when the battery is down, or from physically replacing the latter to get nodes back on track. Clearly, beside acting at the level of consumption, it is pivotal to do something for what concerns energy *procurement*. This is the main objective of energy harvesting (EH) strategies, technologies and devices [57], [58], the latter commonly addressed as energy harvesters (EH). These items are able to sense energy scattered in the surrounding environment under different forms, e.g., mechanical vibrations [59], temperature and thermal gradients [60], [61], light [62], and electromagnetic waves [63], [64], and transduce part of this energy into electricity. EH solutions allow significant room for miniaturization and hybridization [65], and bear the potential to enable fully autonomous self-powered IoT remote smart nodes at the edge [66–69].

2.3.7 Fast and efficient storage and transfer of energy (TMT-7)

The combination of the decreasing trend in power consumption (TMT-5) along with the availability of energy converted from the surrounding environment (TMT-6) marks a significant step toward self-sufficiency of smart nodes at the edge. Yet, something is still missing. In fact, the amount of energy ensured by employing EH devices is subjected to the fluctuations of the corresponding environmental sources. For example, the energy ensured by illumination (solar or artificial), is not available when the device is in the darkness. Similar considerations apply to mechanical vibrations, thermal gradients, etc. A strategy to make the environmental source more reliable is that of diversifying EH addressing orthogonal forms of energy (light, temperature, vibration transducers) within the same HW platform. However, this does not secure edge nodes from potential blackouts.

That said, the availability of more efficient, miniaturized, integrated and long lasting batteries, is crucial and is being effectively pushed by current research [70–72]. In addition, a rather unprecedented solution is also emerging. It has to do with the ability to transfer energy, with real-time on-demand strategies, from one point to another of the edge device, as well as across different remote nodes. This *brokerage-like* approach addresses the need to transfer energy from where it is available, to where it is needed, each time a transaction of this type is necessary, making the whole system more efficient and resilient [73], [74]. At the technical level, the point-to-point transfer of energy is what wireless power transfer (WPT) solutions aim to do [75–77].

2.3.8 Massive simplification of architectural complexity (TMT-8)

Putting together what was discussed to this point, the trend to the increase of *everything* is undeniable. At functional level, operation and functionalities of nodes at the edge are expected to extend, as addressed by TMT-3, while unprecedented physical components, like EH and other redundant items, will commence populating remote nodes. All this looks challenging enough as it is. However, in order to be fully sustainable from different perspectives, including what will be covered by TMT-10 and TMT-11 around complexity, volumes and costs of manufacturing, the increasing scenario of functions and functionalities should be accompanied by a reduction in complexity of the physical systems, sub-systems and platforms implementing them. Meeting all these conflicting demands is possible if unprecedented approaches are embraced, stepping beside and beyond well-established solutions, at any relevant level, e.g., computation [78], design [79], technologies [80], etc. These aspects will be covered with more detail in the continuation of this work.

2.3.9 Incorporation of intelligence at low-complexity device level (TMT-9)

In close proximity to the ongoing discussion, the availability of EI at the functional level requires unavoidably the presence of resident HW components for storing and elaborating data within remote nodes. This has to be intended according to a twofold strategy. On one side, intelligence can be gathered together within a unique physical component (a sort of processor or controller), which is part of the edge node architecture. In addition to this classical approach, a more disruptive and unprecedented approach is expected to unleash the power of edge computing to a much wider extent. The latter has to do with embodying tiny amounts of intelligence within basic HW components, whose nominal function has not necessarily anything to do with treating information. This could be, for instance, the case of an inertial sensor that is also capable of intrinsically performing pre-treatment or partial elaboration of the collected data. The underlying potential of *pulverized* intelligence, scattered down to the simplest HW device, is that of overcoming the classical concept of *system for computation*, with memory, processing/control unit, sensors/actuators for monitoring and reaction purposes, along with ad-hoc SW routines, and so on.

2.3.10 Reduction of complexity and of manufacturing costs (TMT-10)

The TMT that is going to be covered here, along with the next and last TMT-11, transcend all the previous ones, as their focus moves away from functionalities and performance to approach aspects of sustainability from the point of view of manufacturing. Putting together the various points covered by previous TMT, along with the numbers discussed and shown in Figure 2.3, it is clear that thinking to address the resulting manufacturing demands, relying exclusively on standard silicon-based semiconductor processing is simply not an option. On the other hand, availability of lower-cost and simplified technologies will be crucial. To this end, a couple of additional considerations are worth unfolding here. Probably because of the relentless advancement of semiconductor technology according to Moore's law, we are used to referring to state-of-the-art solutions, always looking at the next generation, while not considering the previous ones as still suitable for a wide range of functions. In contrast to this cutting edge-oriented attitude, it must be stressed that a significant part of the small *pieces* of intelligence scattered at the edge do not require ultimate CMOS circuitry to be implemented, both concerning capacity and speed of computation. This opens up to a multi-path scenario. On one hand, past semiconductor technologies with more relaxed features and performance, could be suitable for EI. However, the manufacturing costs, despite lower if compared to state-of-the-art solutions, are likely to be still not fully sustainable, and probably the characteristics of semiconductors still excessive, when compared to what is needed for edge computing. On the other hand, unprecedented non-standard technical solutions can be investigated to yield proper performance and low-cost circuits.

In this sense, the research community has been active for years in developing non-silicon solutions for the realization of semiconductor devices. This is the case of the so-called flexible electronics [81], [82], primarily investigated in the early days to manufacture transistor-based sensors, actuators and circuits onto non-rigid substrates, so they are conformal to uneven surfaces, e.g., human body, garments, wearable items, and so on [83], [84]. In addition to these remarkable characteristics, the technologies for flexible electronics have proven to be a rather low-cost manufacturing processes, also by virtue of them stepping aside classical photolithography-based patterning (e.g., printed electronics) [85–88]. These are certainly crucial aspects in the scenario of EI for 6G and FN. As a final remark, it must be remenbered that the just-sketched scenario can be further boosted by leveraging the possibilities offered by integration solutions (TMT-2), along with the ongoing trend to miniaturization (TMT-1).

2.3.11 Massive increase of manufacturing volumes (TMT-11)

This conclusive TMT is closely related to the previous one, yet, at the same time, is a field of convergence for all the technology-related TMT discussed up to this point. To cut a long story short, the development of flexible and low-cost electronics (i.e., non-standard and non-silicon manufacturing), along with the wide opportunities achievable with integration techniques and solutions, enable the realization of the so-called large area electronics. Such a term groups a plethora of different technical approaches, ranging from patterning based on direct printing, rolling solutions, etc., to die assembly, carrier substrates and various other integration steps, aimed at manufacturing electronics on large areas, thus overcoming the limitations of standard silicon technologies, with particular emphasis on wafer area, yield, size incompatibility of chips, along with processing costs [89–97].

2.4 Deriving (D) Forecasted Functional Features (3F)

Up to this point, the identifying (I) and defining (D) phases of the I2D model are complete. Once again, it is worth stressing the HW centricity of the TMT unfolded in the previous section. That said, it is time to close the loop of this approach, focusing now on the third and last deriving (D) phase. Ideally, the latter one aims to define the vertical dimension in the conceptual space shown in Figure 2.1. In practical terms, the 2D plane framed by CMR and TMT defines a plethora of different and multiple combinations of such elements, whose mixing gives rise to factual characteristics, which are expected to be scored by future HW components. These desirable performances are termed here as forecasted functional features (3F), and a few of them are going to be analyzed below as reference examples. In fact, 3F are conceptually adjacent to device specifications, therefore examples of physical HW components can be easily linked to them. To this end, it must be stressed that for each of the 3F listed below, there exist entries in the scientific literature discussing similar concepts. This is a clear indication that the concepts developed in this work, although rather unprecedented, are aligned to the possibilities offered by state-of-the-art miniaturized HW technologies.

2.4.1 Multi-functional basic HW devices (3F-1)

The first 3F here at stake is the ability of a unique basic HW device to implement diverse and orthogonal functions. To this end, a reference example is discussed in [98] and also reported in [25]. The device is the classical mass–spring system shown schematically in Figure 2.4.

It features fixed and movable interdigitated (comb-like) finger electrodes, to transduce a mechanical displacement, induced by an acceleration, into a variation of capacitance. This type of sensor geometry and its working principle are well known and have been exploited for decades in commercial inertial

Figure 2.4: Schematic of the mass–spring interdigitated capacitive transduction inertial sensor in (a) its rest position, and (b) when subjected to an acceleration of amplitude a.

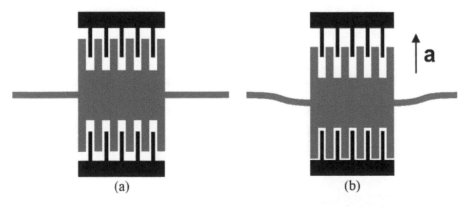

(a)   (b)

sensors. However, the way it is exploited in [98] bears significant innovation. In fact, a unique (MEMS-based) inertial device is operated as an ambient pressure sensor, an environmental temperature sensor, and is as well as an accelerometer and gyroscope. The work also covers a discussion around the development of ad-hoc control electronics, which enable operating the micro-transducer in all the different sensing configurations mentioned above.

2.4.2 Multi-functional monolithic HW devices (3F-2)

This 3F is conceptually similar to 3F-1, as the aim is to implement different sensing and transducing functionalities, although a substantial difference separates the two 3F. In fact, in this case the focus is on monolithic realizations of multiple devices. In other words, 3F-1 addresses a unique HW item that can work in different ways, while 3F-2 targets sets of different HW items, manufactured in a unique technology and highly miniaturized.

In light of the above discussion, a relevant contribution to be reported is available in [99]. In it, a micro-fabrication processing technology platform is discussed, and incorporates a mix of fabrication steps belonging both to bulk and surface micromachining (see Section 1.2.3). The CMOS-compatible process flow enables the monolithic manufacturing of five different sensors, within a 3×3 mm^2 silicon chip. The mentioned devices are temperature, corrosion, relative humidity, gas, and gas flow sensors. The schematic cross-section of the technology is shown in Figure 2.5.

Other contributions exploiting similar approaches are also available in [100–104], demonstrating the wide functional flexibility offered by micro-fabrication technologies.

Figure 2.5: Schematic cross-section of the monolithically manufactured five sensors reported in [99].

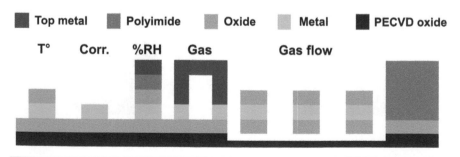

2.4.3 Multi-functional in-package integrated HW devices (3F-3)

As already mentioned, packaging and integration solutions offer wide possibilities in terms of integration of heterogeneous devices realized in incompatible technologies, them being a valuable alternative to monolithic manufacturing. A common packaging technique consists in exploiting an entire wafer as a package, the latter of the same size as the device wafer. It is called WLP, and a sketch is offered in Figure 2.6.

Figure 2.6: (a) W2W alignment of the package to the device wafer. (b) Details of TWV interconnects to redistribute electrical signals from the device to the external world.

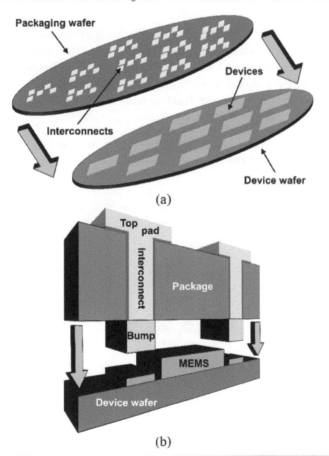

The package is provided with sets of electrical interconnects that are placed in correspondence to signal pads on the device wafer. During the W2W alignment, the package contacts electrically the device layer, as reported in Figure 2.6 (a). A closer view of what happens at device level is available in Figure 2.6 (b). TWV across the package wafer are first etched and then filled with conductive material. This way, the electrical signals are redistributed from the packaged devices (on the device wafer), to the external world, i.e., on top of the package wafer. The physical contact of bumps on the package side, to pads on the device side, ensure both electrical continuity and mechanical adhesion between the two wafers. That said, significant examples of SiP are available in the literature, e.g., as discussed in [105–110].

2.4.4 Basic HW devices for harvesting of energy from the environment (3F-4)

It was outlined before that EH is a key-technology to address distributed power demands of EI and remote nodes. The attention here is focused on physical implementations of such energy converters. For completeness, beside conversion (i.e., EH), distribution (i.e., WPT) and storage (i.e., batteries) of energy are also covered by this 3F. Having said that, the main classes of devices are schematically reported in Figure 2.7.

The scientific literature is populated by a significant number of contributions for each of the classes of devices listed in Figure 2.7.

Figure 2.7: Examples of EH and energy storage/distribution for what concerns: (a) mechanical vibrations; (b) thermal gradients; (c) light; (d) electromagnetic waves; (e) WPT; (f) thin-film miniaturized batteries.

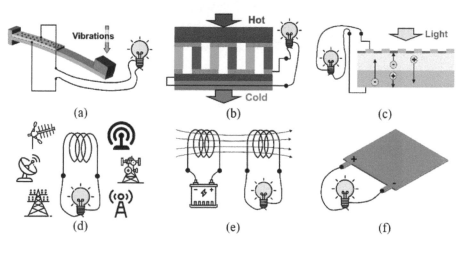

Starting from the conversion of environmental mechanical vibrations (Figure 2.7 (a)), many published works deal with piezoelectric [111–113], electrostatic (capacitive) [114], [115], and magnetic transduction [116], [117]. Multiple transduction mechanisms in the same device, as well as design concepts to extend the frequency range of EH operability, are also covered [118–120].

When thermal gradients are concerned (Figure 2.7 (b)), there are significant examples of EH transducers exploiting miniaturized thermocouples in semiconductor technologies [121–123], along with more than Moore technologies that use, for instance, CNT [124–126].

Moving on to light as an energy source (Figure 2.7 (c)), in spite of the maturity of photovoltaic cells technology, emerging distributed IoT-like applications are driving research and development of miniaturized EH converters [127–130].

Addressing now electromagnetic radiation scattered in the environment (Figure 2.7 (d)), the focus of research on EH solutions targets the development and optimization both of antennas for capturing part of the radiated power, along with electronic circuits for efficient conversion of RF waves into DC currents [131–137].

Differently from EH converters, in WPT systems energy is intentionally *transmitted* from an active (Tx) to a passive (Rx) end (Figure 2.7 (e)), both via RF or inductive coupling [138]. That said, many aspects are under investigation, e.g., efficiency, miniaturization, integration, and so on [139–141].

Stepping to the aspect of energy storage, the research is very active in developing miniaturized batteries (Figure 2.7 (f)). To this end, thin-films based on typical semiconductor technologies are investigated [142], [143], along with the employment of flexible substrates [144].

Eventually, a final consideration is worth mentioning. It has already been stressed that a limitation of environmental EH is linked to the scarcely predicable variability of energy, e.g., when light is turned off, or when a certain mechanical vibration suddenly stops. An effective strategy to overcome these constraints is to diversify the types of EH converters (e.g., vibration, thermal, electromagnetic, etc.), putting them together within a unique HW platform. To this end, leveraging the wide possibilities offered by miniaturization and integration, SiP featuring energy converters, along with batteries, can be assembled together [145–150].

2.4.5 Sensors and actuators able to autonomously harvest energy (3F-5)

The previous 3F-4 offered a wide perspective of possibilities in terms of physical miniaturized devices for EH, WPT and energy storage. The further step achieved by the current 3F-5 is that of incorporating EH functionalities within physical sensors or transducers. This means moving in the direction of self-powered basic HW components.

To this end, a significant example available in literature is a miniaturized tri-axial inertial sensor, featuring a seismic cylindrical mass and four suspending slender beams. The schematic is reported in Figure 2.8 and its technical details are thoroughly discussed in [151–153].

The device is miniaturized to a certain extent, albeit it is not manufactured by means of micro-fabrication technologies. Its dimensions are around $110 \times 110 \times 20$ mm^3, but, it admits significant room for being further miniaturized and realized in MEMS technology. The cornerstone technical characteristic of the design concept is the concurrent exploitation of piezoelectric and

Figure 2.8: Schematic of the self-powered miniaturized tri-axial accelerometer discussed in [151]–[153].

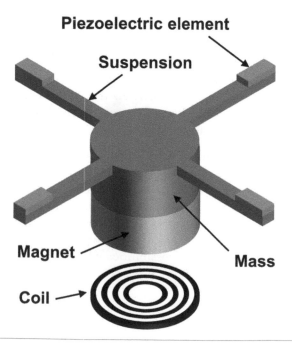

electromagnetic transduction mechanisms. In detail, the former serves the sensing of acceleration, transducing mechanical deformations into proportional amounts of electric charges. The latter, instead, operates the harvesting of energy consequent to the mechanical deformation of the mass–spring system. The energy extracted from vibrations powers the charge amplifier that reads the accelerometer output signal, along with the control circuitry connected to the inertial sensors. The latter also compensates for the presence of the electromagnetic EH, as it unavoidably influences the inertial sensor dynamic behavior, thus yielding a self-powered sensing HW sub-system. Preliminary simulated data report that the EH part of the device is able to provide 10 mW under a horizontal vibration at 100 Hz.

2.4.6 Self-reacting sensors and actuators (3F-6)

The 3F at stake here covers the ability of basic HW components to self-react to certain surrounding conditions, as soon as the latter ones occur, thus making the physical item itself work as if it was a more complex sub-system. These features are crucial in a scenario of pronounced pervasivity of functions and intelligence at the edge, as it can enable significant simplification at HW level. The examples that are going to be reported focus on self-healing devices. However, it must be noted that self-reactive characteristics include also other aspects, like reconfiguration, restoration, etc., not detailed here for the sake of brevity [154].

Given these considerations, the work in [155] reports on the phenomenon of electrical treeing in insulating materials subjected to prolonged high-voltage drops (implying very-intense electric field). The resulting mechanical cracks typically initiate from materials imperfections, e.g., voids and steps across the material over time. When the tree-like cracks reach the boundaries of the insulating material, they often lead to dielectric breakdown, which is an irreversible malfunctioning condition. The problem is counteracted by adding micro-capsules within the polymer-based dielectric, as shown in Figure 2.9.

Such capsules are filled with a liquid healing agent, and are scattered within the insulating material. When these reservoirs break, their content is released, thus restoring the material properties. The sustained hypothesis in [155] is that treeing itself releases the healing agent, when cracks reach and break one or more micro-capsules membranes. The released monomer fills the cracks and heat, due to normal operation of the dielectric, triggers the polymerization that completes the healing process.

Figure 2.9: Schematic of the self-healing dielectric provided with healing agent capsules, broken in case of treeing [155].

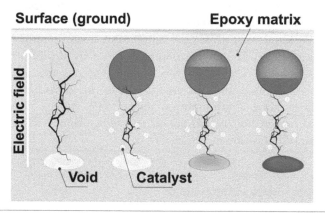

Other scientific works cover the exploitation of micro-capsules to trigger self-healing and self-repairing features in other fields, like in cementitious materials [156], characterized by diverse activation mechanisms, such as ultrasonic waves [157].

Scientists also focus their attention on materials that implement self-healing properties employing vascular-type structures, inspired by human body and plants. Differently from the case of micro-capsules, in this case a network of micro-channels is deployed at the design level, and when the need to restore a certain property of the material occurs, healing agents are pumped through the vascular network, mimicking blood, lymph and sap [158–160].

In addition to the rather complex structures mentioned earlier, materials with intrinsic self-healing properties are also reported in literature. The work in [161] discusses restoration of the mechanical resonant properties of cubic SiC and Ge thin-films. Such 3D layers are intentionally altered by ultrasonic excitation and temperature cycling stress, and then recovered by means of heating.

Eventually, a comprehensive review of innovative materials exhibiting self-healing characteristics, with focus on high electrical stress applications, is provided in [162].

2.4.7 Sensors and actuators embodying conditioning and/or elaboration (3F-7)

This last 3F is conceptually similar to the previous one, as the implementation of orthogonal functions by a unique low-complexity HW component is at stake.

In this case, intrinsic intelligence is analyzed. In other words, 3F-7 outlines the capacity of a basic sensor or actuator to perform also (elementary) conditioning and/or elaboration of the transduced magnitude.

The example chosen here is that of the MEMS-based inertial sensor reported in [163], and shown schematically in Figure 2.10.

Figure 2.10: Schematic of the 3-bit accelerometer in [163], based on coupled switches and binary output, with reference to (a) bit 0 (LSB); (b) bit 1; (c) bit 2 (MSB).

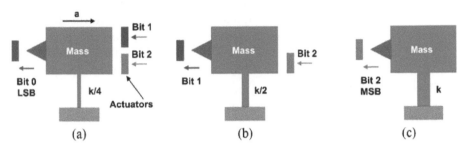

The design concept realizes a self-computing digital accelerometer, which makes redout electronics redundant and (to a certain extent) unnecessary. This is made possible by cascading as many inertial MEMS ohmic switches as the desired resolution of the sensor in terms of bits. Each micro-relay proof-mass is connected to a deformable suspension with different elastic constant. In the 3-bit simplified context in Figure 2.10, the elastic constant k is varied acting on the width of the deformable bar keeping the seismic mass suspended. In particular, it is set to $k/4$ for bit 0 (LSB), to $k/2$ for bit 1, and to k for bit 2 (MSB). The output configuration that corresponds to a certain acceleration amplitude is obtained by cascading the voltage drops controlled by each inertial switch activation, among a set of other micro-relays composing the inertial sensor. This way, the inertial sensor is able to provide itself a discretized (digital) output voltage per each acceleration level sensed, without the need of ad-hoc readout electronic circuitry performing such a conditioning function.

2.5 Wrap-up of the I2D model and beyond

The previous pages developed an extensive discussion around the I2D model along with its constitutive dimensions. The aim of this conclusive section is twofold. First, a brief summary of such a conceptual tool is going to be covered, stressing the intricacy of links existing among its elements. This is done in a quantitative-like fashion, by reporting correlation maps, showing, per each 3F, the extent of dependency between CMR and TMT. Then, an outlook on what can be triggered by the I2D model is going to be provided.

2.5.1 Maps of correlation between the three dimensions of the I2D model

In the following, the set of correlation links between the three dimensions of the I2D model is going to be unrolled. Recalling Figure 2.1, the target of this section is that of providing a more quantitative insight around the 3D space, having CMR, TMT and 3F as axes. First, a recap of the meaning of each CMR, TMT and 3F is developed Table 2.1, Table 2.2, and Table 2.3, respectively. This will help keeping such a nomenclature readily at hand during the upcoming discussion.

Table 2.1: Recap of CMR (see Section 2.2 for more details).

CMR-1	Volume of data over the network
CMR-2	Centralization vs. distribution of intelligence
CMR-3	Edge intelligence (EI) and local self-management
CMR-4	Sensing and conditioning at the edge
CMR-5	Data storage and computation at the edge
CMR-6	Energy distribution for edge operation
CMR-7	Orchestration of a network of networks

That said, the correlation is going to be developed in a dedicated plot per each of the seven 3F. In particular, the chosen plot scheme reports the 11 TMT on the horizontal axis, and the seven CMR on the vertical axis. The correlation plots for the 3F, from 3F-1 to 3F-7, are collected in the following images, from Figure 2.11 to Figure 2.17, respectively.

For the sake of brevity, the task of unfolding further a discussion on the data reported above is left to the reader. The only aspect worthwhile stressing is the complete coverage of the plot area with high levels of correlation when collectively looking at the set of seven 3F.

Table 2.2: Recap of TMT (see Section 2.3 for more details).

TMT-1	Increase of miniaturization
TMT-2	Increase of integration
TMT-3	Increase of operation differentiation and diversity
TMT-4	Increase of redundancy
TMT-5	Decrease of power consumption
TMT-6	Harvesting of energy from the environment
TMT-7	Fast and efficient storage and transfer of energy
TMT-8	Massive simplification of architectural complexity
TMT-9	Incorporation of intelligence at low-complexity device level
TMT-10	Reduction of complexity and of manufacturing costs
TMT-11	Massive increase of manufacturing volumes

Table 2.3: Recap of 3F (see Section 2.4 for more details).

3F-1	Multi-functional basic HW devices
3F-2	Multi-functional monolithic HW devices
3F-3	Multi-functional in-package integrated HW devices
3F-4	Basic HW devices for harvesting of energy from the environment
3F-5	Sensors and actuators able to autonomously harvest energy
3F-6	Self-reacting sensors and actuators
3F-7	Sensors and actuators embodying conditioning and/or elaboration

Figure 2.11: Correlation map of the I2D dimensions for 3F-1.

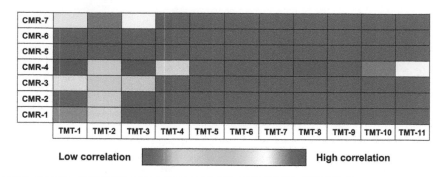

Figure 2.12: Correlation map of the I2D dimensions for 3F-2.

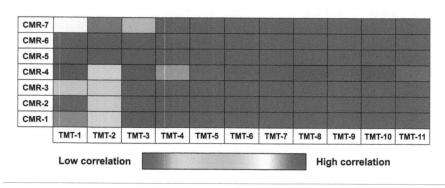

Figure 2.13: Correlation map of the I2D dimensions for 3F-3.

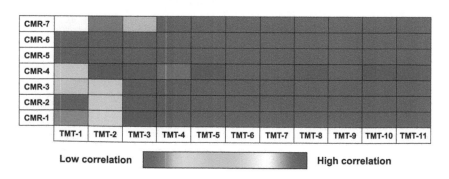

Figure 2.14: Correlation map of the I2D dimensions for 3F-4.

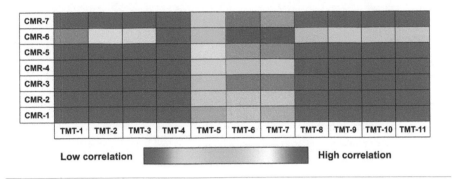

Figure 2.15: Correlation map of the I2D dimensions for 3F-5.

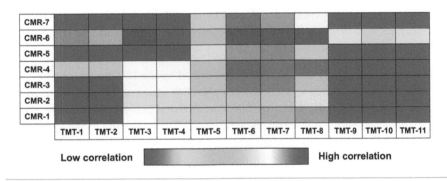

Figure 2.16: Correlation map of the I2D dimensions for 3F-6.

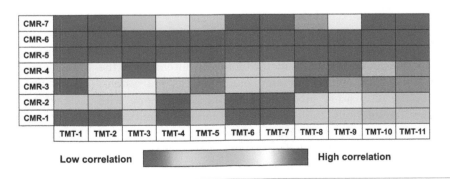

Figure 2.17: Correlation map of the I2D dimensions for 3F-7.

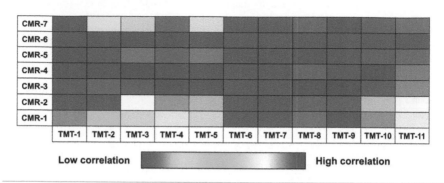

2.5.2 Beyond the I2D model

Taking the base of knowledge developed by the I2D model as reference ground, this section will develop a few considerations on what could be built on it, looking at EI in the frame of 6G and FN.

Putting together all the elements introduced by the I2D model, and keeping the discussion as general as possible, the HW must gain and incorporate features typical of the SW, like flexibility, adaptivity, modularity, self-diagnostic capacities, and, more in general, intelligence, thus becoming more symmetric with respect to SW [164]. Augmented HW–SW symmetry would increase their separation, triggering a twofold opposed effect. On one hand, the well-established HW–SW co-design will be overtaken, loosening the mutual optimization of the involved (sub-)system. Conversely, this would unleash dynamic aggregation of HW and SW modules, empowering the evolutionary and self-managing (AI-boosted) characteristics of the network edge (i.e., EI), mentioned above [164].

Bearing in mind such a context, it is straightforward that HW technologies at the edge are pivotal. To this end, a crucial role is envisioned for the broad area of microtechnologies and nanotechnologies, intended in the widest possible way. This means embodying micro-/nano-devices and systems (MEMS/NEMS), materials, fabrication processes and platforms, along with packaging and integration solutions [165]. To this end, it has to be mentioned that in the ample discussion around the I2D model, the centrality of MEMS/NEMS technology was never mentioned explicitly. This is because the followed approach is not technology-push- like, but rather driven by emerging needs and requirements.

Figure 2.18: The concept of SiLCHI supported by the seven 3F.

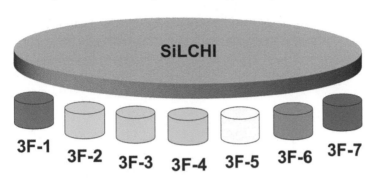

Eventually, the items identified by the I2D model are certainly diverse and comprehensive. However, such a description must not be regarded as rigid, since many extensions are possible, and will certainly be necessary, as much as the remaining distance from 6G and FN will progressively be covered.

To conclude this chapter, an example of what can be enabled at a conceptual level by the I2D model is schematically depicted in Figure 2.18.

In this representation, the seven 3F are acting as pillars in sustaining the higher-level concept, here named SiLCHI, which stands for system in a low complexity hardware item. As already discussed, regardless of the specific 3F at stake, the extension and diversification of the functionalities implemented by a certain miniaturized HW item are always addressed. Putting together these features, the comprehensive new concept of basic physical devices that gather together the functions (i.e., SiLCHI) typically implemented by a complete (traditional) sub-system, is envisioned as a key-element in EI of 6G and FN.

2.6 Summary

The main target of this chapter was the deployment of a conceptual framework to support the reformulation of the classical meaning of HW, with specific focus on low-complexity devices for employment at the network edge of 6G and FN. This was done via the discussion of the I2D model, featuring CMR, TMT and 3F. To this end, seven CMR were identified, including, among the others, EI and energy distribution. Diversely, the set of TMT is composed of 11 entries, among which increase of miniaturization/integration, decrease of power consumption and incorporation of intelligence at low-complexity device level were reported. Finally, the set of 3F featured seven entries, among which multi-functional basic/monolithic HW devices and self-reacting sensors/actuators stand. In addition, for each 3F, examples taken from the current literature were reported on devices, systems and solutions that satisfy, at least partially, the identified demands. In particular, for each of the seven 3F the correlation with the bi-dimensional crossing of the seven CMR and 11 TMT was graphically visualized.

References

[1] T. Norp, "5G Requirements and Key Performance Indicators," *Journal of ICT Standardization*, vol. 6, no. 1-2, 2018, doi: 10.13052/jicts2245-800x.612

[2] A. Mourad, R. Yang, P. H. Lehne, and A. De La Oliva, "Towards 6G: Evolution of key performance indicators and technology trends," 2020, doi: 10.1109/6GSUMMIT49458.2020.9083759

[3] *Peak data rates of 4G, 5G, and 6G mobile broadband technologies.* Accessed: Jun. 1, 2023. [Online]. Available: https://www.statista.com/statistics/1183654/mobile-broadband-peak-data-rates/

[4] R. Mijumbi, J. Serrat, J. L. Gorricho, N. Bouten, F. De Turck, and R. Boutaba, "Network function virtualization: State-of-the-art and research challenges," *IEEE Commun. Surv. Tutorials*, vol. 18, no. 1, 2016, doi: 10.1109/COMST.2015.2477041

[5] X. You *et al.*, "Towards 6G wireless communication networks: vision, enabling technologies, and new paradigm shifts," *Science China Information Sciences*, vol. 64, no. 1, 2021, doi: 10.1007/s11432-020-2955-6

[6] *Mobile data traffic in Western Europe from 2016 to 2021.* Accessed: Jun. 1, 2023. [Online]. Available: https://www.statista.com/statistics/292864/western-europe-mobile-data-traffic/

[7] *Quarterly data traffic in mobile broadband networks in Spain from 1st quarter 2013 to 4th quarter 2021.* Accessed: Jun. 1, 2023. [Online]. Available: https://www.statista.com/statistics/462736/quarterly-mobile-broadband-data-traffic-spain/

[8] *6G - Statistics & Facts.* Accessed: Jun. 6, 2023. [Online]. Available: https://www.statista.com/topics/7163/6g/#topicOverview

[9] *Connection density of 4G, 5G, and 6G mobile broadband technologies.* Accessed: Jun. 6, 2023. [Online]. Available: https://www.statista.com/statistics/1183690/mobile-broadband-connection-density/

[10] D. C. Nguyen *et al.*, "6G internet of Things: A Comprehensive Survey," *IEEE internet Things J.*, vol. 9, no. 1, 2022, doi: 10.1109/JIOT.2021.3103320

[11] H. Hu and C. Jiang, "Edge Intelligence: Challenges and Opportunities," 2020, doi: 10.1109/CITS49457.2020.9232575

[12] Z. Zhou, X. Chen, E. Li, L. Zeng, K. Luo, and J. Zhang, "Edge Intelligence: Paving the Last Mile of Artificial Intelligence With Edge Computing," *Proc. IEEE*, 2019, doi: 10.1109/JPROC.2019.2918951

[13] D. Xu *et al.*, "Edge Intelligence: Empowering Intelligence to the Edge of Network," *Proc. IEEE*, vol. 109, no. 11, 2021, doi: 10.1109/JPROC.2021.3119950

[14] *State of the Edge Report 2020*. Accessed: Jun. 7, 2023. [Online]. Available: https://stateoftheedge.com/reports/state-of-the-edge-2020/

[15] *The Emergent Third Act Of The internet: It's Time To Invest*. Accessed: Jun. 7, 2023. [Online]. Available: https://www.forbes.com/sites/forbestechcouncil/2020/07/03/the-emergent-third-act-of-the-internet-its-time-to-invest/?sh=7b82081f9802

[16] *Edge Computing - statistics & facts*. Accessed: Jun. 7, 2023. [Online]. Available: https://www.statista.com/topics/6173/edge-computing/#topicOverview

[17] *13 kinds of sensors in mobile phones and what are recorded by the sensors*. Accessed: Jun. 7, 2023. [Online]. Available: https://www.utmel.com/blog/categories/sensors/13-kinds-of-sensors-in-mobile-phones-and-what-are-recorded-by-the-sensors

[18] V. Taraate, *ASIC Design and Synthesis*. 2021

[19] M. Meribout, A. Baobaid, M. O. Khaoua, V. K. Tiwari, and J. P. Pena, "State of Art IoT and Edge Embedded Systems for Real-Time Machine Vision Applications," *IEEE Access*, vol. 10, 2022, doi: 10.1109/ACCESS.2022.3175496

[20] X. Chen, Y. Cai, Q. Shi, M. Zhao, B. Champagne, and L. Hanzo, "Efficient Resource Allocation for Relay-Assisted Computation Offloading in Mobile-Edge Computing," *IEEE internet Things J.*, vol. 7, no. 3, 2020, doi: 10.1109/JIOT.2019.2957728

[21] N. N. Ei, M. Alsenwi, Y. K. Tun, Z. Han, and C. S. Hong, "Energy-Efficient Resource Allocation in Multi-UAV-Assisted Two-Stage Edge Computing for Beyond 5G Networks," *IEEE Trans. Intell. Transp. Syst.*, 2022, doi: 10.1109/TITS.2022.3150176

[22] X. Yu, K. Yu, X. Huang, X. Dang, K. Wang, and J. Cai, "Computation Efficiency Optimization for RIS-Assisted Millimeter-Wave Mobile Edge Computing Systems," *IEEE Trans. Commun.*, vol. 70, no. 8, 2022, doi: 10.1109/TCOMM.2022.3181673

[23] M. Yao, L. Chen, T. Liu, and J. Wu, "Energy efficient cooperative edge computing with multi-source multi-relay devices," 2019, doi: 10.1109/HPCC/SmartCity/DSS.2019.00126

[24] J. Park and Y. Lim, "Toward Adaptive Energy Management for Mobile Edge Networks," 2022, doi: 10.1109/ECICE55674.2022.10042829

[25]   J. Iannacci and H. V. Poor, "Review and Perspectives of Micro/Nano Technologies as Key-Enablers of 6G," *IEEE Access*, vol. 10, 2022, doi: 10.1109/ACCESS.2022.3176348

[26]   A. Kunnappilly, P. Backeman, and C. Seceleanu, "UML-based modeling and analysis of 5G service orchestration," in *Proceedings - Asia-Pacific Software Engineering Conference, APSEC*, 2020, vol. 2020-December, doi: 10.1109/APSEC51365.2020.00021

[27]   S. Aleyadeh, A. Moubayed, and A. Shami, "Mobility Aware Edge Computing Segmentation Towards Localized Orchestration," 2021, doi: 10.1109/ISNCC52172.2021.9615795

[28]   L. Liu, "SDN orchestration for dynamic end-to-end control of data center multi-domain optical networking," *China Communications*, vol. 12, no. 8. 2015, doi: 10.1109/CC.2015.7224702

[29]   H. Baba *et al.*, "End-to-end 5G network slice resource management and orchestration architecture," 2022, doi: 10.1109/NetSoft54395.2022.9844088

[30]   I. Badmus, A. Laghrissi, M. Matinmikko-Blue, and A. Pouttu, "End-to-end network slice architecture and distribution across 5G micro-operator leveraging multi-domain and multi-tenancy," *Eurasip J. Wirel. Commun. Netw.*, vol. 2021, no. 1, 2021, doi: 10.1186/s13638-021-01959-7

[31]   G. Baggio, A. Francescon, and R. Fedrizzi, "Multi-domain service orchestration with X-MANO," 2017, doi: 10.1109/NETSOFT.2017.8004259

[32]   J. Liao *et al.*, "Design and Implementation of Heterogeneous Orchestration for 5G's Scenarios," 2021, doi: 10.1109/ICFTIC54370.2021.9647301

[33]   X. Chen, H. Zeng, and T. Wu, "Decentralized orchestration with local centralized orchestration for composite Web services," 2010, doi: 10.1109/PDCAT.2010.16

[34]   A. Mayoral *et al.*, "Control orchestration protocol: Unified transport API for distributed cloud and network orchestration," *J. Opt. Commun. Netw.*, vol. 9, no. 2, 2017, doi: 10.1364/JOCN.9.00A216

[35]   W. A. Atherton, *From Compass to Computer*. 1984

[36]   R. R. Schaller, "Moore's law: past, present, and future," *IEEE Spectr.*, vol. 34, no. 6, 1997, doi: 10.1109/6.591665

[37]   A. B. Kahng, "Scaling: More than Moore's law," *IEEE Des. Test Comput.*, vol. 27, no. 3, 2010, doi: 10.1109/MDT.2010.71

[38]   C. M. N. Brigante, N. Abbate, A. Basile, A. C. Faulisi, and S. Sessa, "Towards miniaturization of a MEMS-based wearable motion capture system," *IEEE Trans. Ind. Electron.*, vol. 58, no. 8, 2011, doi: 10.1109/TIE.2011.2148671

[39]   J. M. Early, "Semiconductor Devices," *Proc. IRE*, vol. 50, no. 5, 1962, doi: 10.1109/JRPROC.1962.288388

[40]   K. Merghem, S. F. Busch, F. Lelarge, M. Koch, A. Ramdane, and J. C. Balzer, "Terahertz Time-Domain Spectroscopy System Driven by a Monolithic Semiconductor Laser," *J. Infrared, Millimeter, Terahertz Waves*, vol. 38, no. 8, 2017, doi: 10.1007/s10762-017-0401-2

[41]   J. H. Lau, *Heterogeneous integrations*. 2019

[42]   K. Tu, C. Chen, and H. Chen, *Electronic Packaging Science and Technology*. 2021

[43]   P. Garrau, P.; Bower, C.; Ramm, *Handbook of 3D Integration Vol.1.pdf*, vol. 3. 2008

[44]   S. K. Lim, *Design for High Performance, Low Power, and Reliable 3D Integrated Circuits*. 2013

[45]   I. A. M. Elfadel and G. Fettweis, *3D stacked chips: From emerging processes to heterogeneous systems*. 2016

[46]   F. Santagata *et al.*, "System in package (SiP) technology: fundamentals, design and applications," *Microelectron. Int.*, vol. 35, no. 4, 2018, doi: 10.1108/MI-09-2017-0045

[47]   S. Santhanamahalingam, S. Alagarsamy, and K. Subramanian, "A study of cloud-based VPN establishment using network function virtualization technique," 2022, doi: 10.1109/ICOSEC54921.2022.9951894

[48]   M. H. Tsai, H. T. Liang, Y. H. Wang, and W. C. Chung, "Enhanced OpenStack Cloud for Network Function Virtualization," 2020, doi: 10.1109/ICS51289.2020.00045

[49]   J. Kang, J. Kang, and O. Simeone, "On the Trade-Off between Computational Load and Reliability for Network Function Virtualization," *IEEE Commun. Lett.*, vol. 21, no. 8, 2017, doi: 10.1109/LCOMM.2017.2698040

[50]   G. Kun *et al.*, "'Opened' or 'Closed' RAN in 5G," 2022, doi: 10.1109/SAMI54271.2022.9780667

[51]   S. K. Singh, R. Singh, and B. Kumbhani, "The Evolution of Radio Access Network Towards Open-RAN: Challenges and Opportunities," 2020, doi: 10.1109/WCNCW48565.2020.9124820

[52]   A. De Javel, J. S. Gomez, P. Martins, J. L. Rougier, and P. Nivaggioli, "Towards a new open-source 5G development framework: An introduction to free5GRAN," in *IEEE Vehicular Technology Conference*, 2021, vol. 2021-April, doi: 10.1109/VTC2021-Spring51267.2021.9448964

[53]   J. S. Lee, J. Park, J. Choi, and M. S. Lee, "Design of a Management Plane for 5G Open Fronthaul Interface," in *International Conference on ICT Convergence*, 2020, vol. 2020-October, doi: 10.1109/ICTC49870.2020.9289612

[54]   T. Hiramoto *et al.*, "Ultra-low power and ultra-low voltage devices and circuits for IoT applications," 2016, doi: 10.1109/SNW.2016.7578025

[55] Y. Pu *et al.*, "Blackghost 1.0 test chip: On the road towards commercializing ultra-low-Vdd SoC for internet-of-Things," in *2017 IEEE SOI-3D-Subthreshold Microelectronics Unified Conference, S3S 2017*, 2018, vol. 2018-March, doi: 10.1109/S3S.2017.8308750

[56] B. Y. Chang and C. F. Jou, "Design of a 3.1-10.6GHz low-voltage, low-power CMOS low-noise amplifier for ultra-wideband receivers," in *Asia-Pacific Microwave Conference Proceedings, APMC*, 2005, vol. 2, doi: 10.1109/APMC.2005.1606458

[57] S. Priya, D. J. Inman. *Energy Harvesting Technologies*. 2009

[58] Y. K. Tan. *Sustainable Energy Harvesting Technologies - Past, Present and Future*. 2011

[59] P. Gljuši, S. Zelenika, D. Blaževi, and E. Kamenar, "Kinetic energy harvesting for wearable medical sensors," *Sensors (Switzerland)*, vol. 19, no. 22, 2019, doi: 10.3390/s19224922

[60] D. Zabek and F. Morini, "Solid state generators and energy harvesters for waste heat recovery and thermal energy harvesting," *Thermal Science and Engineering Progress*, vol. 9. 2019, doi: 10.1016/j.tsep.2018.11.011

[61] R. A. Kishore and S. Priya, "A Review on low-grade thermal energy harvesting: Materials, methods and devices," *Materials*, vol. 11, no. 8. 2018, doi: 10.3390/ma11081433

[62] H. Jabbar and T. Jeong, "Ambient Light Energy Harvesting and Numerical Modeling of Non-Linear Phenomena," *Appl. Sci.*, vol. 12, no. 4, Feb. 2022, doi: 10.3390/app12042068

[63] H. Lee and J. S. Roh, "Charging device for wearable electromagnetic energy-harvesting textiles," *Fash. Text.*, vol. 8, no. 1, 2021, doi: 10.1186/s40691-020-00233-6

[64] M. Gholikhani, S. A. Tahami, M. Khalili, and S. Dessouky, "Electromagnetic energy harvesting technology: Key to sustainability in transportation systems," *Sustain.*, vol. 11, no. 18, 2019, doi: 10.3390/su11184906

[65] H. Liu, H. Fu, L. Sun, C. Lee, and E. M. Yeatman, "Hybrid energy harvesting technology: From materials, structural design, system integration to applications," *Renewable and Sustainable Energy Reviews*, vol. 137. Elsevier Ltd, Mar. 01, 2021, doi: 10.1016/j.rser.2020.110473

[66] A. D. Ball, F. Gu, R. Cattley, X. Wang, and X. Tang, "Energy harvesting technologies for achieving self-powered wireless sensor networks in machine condition monitoring: A review," *Sensors (Switzerland)*, vol. 18, no. 12. MDPI AG, 2018, doi: 10.3390/s18124113

[67] A. J. Williams, M. F. Torquato, I. M. Cameron, A. A. Fahmy, and J. Sienz, "Survey of Energy Harvesting Technologies for Wireless Sensor Networks," *IEEE Access*, vol. 9, pp. 77493–77510, 2021, doi: 10.1109/ACCESS.2021.3083697

[68]   M. Shirvanimoghaddam *et al.*, "Towards a Green and Self-Powered internet of Things Using Piezoelectric Energy Harvesting," *IEEE Access*, vol. 7, 2019, doi: 10.1109/ACCESS.2019.2928523

[69]   T. Sanislav, G. D. Mois, S. Zeadally, and S. C. Folea, "Energy Harvesting Techniques for internet of Things (IoT)," *IEEE Access*, vol. 9, 2021, doi: 10.1109/ACCESS.2021.3064066

[70]   T. Wu, W. Dai, M. Ke, Q. Huang, and L. Lu, "All-Solid-State Thin Film μ-Batteries for Microelectronics," *Advanced Science*, vol. 8, no. 19. 2021, doi: 10.1002/advs.202100774

[71]   A. Rambabu, S. B. Krupanidhi, and P. Barpanda, "An overview of nanostructured Li-based thin film micro-batteries," *Proceedings of the Indian National Science Academy*, vol. 85, no. 1. 2019, doi: 10.16943/ptinsa/2018/49472

[72]   N. J. Dudney, "Thin film micro-batteries," *Electrochem. Soc. Interface*, vol. 17, no. 3, 2008, doi: 10.1149/2.f04083if

[73]   R. J. Tom, S. Sankaranarayanan, and J. J. P. C. Rodrigues, "Agent negotiation in an IoT-Fog based power distribution system for demand reduction," *Sustain. Energy Technol. Assessments*, vol. 38, 2020, doi: 10.1016/j.seta.2020.100653

[74]   A. Bregar, "Implementation of a multi-agent multi-criteria negotiation protocol for self-sustainable smart grids," *J. Decis. Syst.*, vol. 29, no. sup1, 2020, doi: 10.1080/12460125.2020.1848374

[75]   Z. Zhang and H. Pang, *Wireless Power Transfer: Principles and Applications*. 2022

[76]   L. Wang, U. K. Madawala, J. Zhang, and M. C. Wong, "A New Bidirectional Wireless Power Transfer Topology," *IEEE Trans. Ind. Appl.*, vol. 58, no. 1, 2022, doi: 10.1109/TIA.2021.3097015

[77]   O. Okoyeigbo, A. A. Olajube, O. Shobayo, A. Aligbe, and A. E. Ibhaze, "Wireless power transfer: A review," in *IOP Conference Series: Earth and Environmental Science*, 2021, vol. 655, no. 1, doi: 10.1088/1755-1315/655/1/012032

[78]   E. van Krieken, E. Acar, and F. van Harmelen, "Analyzing Differentiable Fuzzy Logic Operators," *Artif. Intell.*, vol. 302, Jan. 2022, doi: 10.1016/j.artint.2021.103602

[79]   A. Thompson, P. Layzell, and R. S. Zebulum, "Explorations in design space: Unconventional electronics design through artificial evolution," *IEEE Trans. Evol. Comput.*, vol. 3, no. 3, 1999, doi: 10.1109/4235.788489

[80]   J. Song, "Editorial: Recent advances in mechanics of unconventional electronics," *Theoretical and Applied Mechanics Letters*, vol. 6, no. 1. 2016, doi: 10.1016/j.taml.2016.02.001

[81]   V. K. Khanna, *Flexible Electronics, Volume 1 – Mechanical background, materials and manufacturing*. 2019

[82]   V. K. Khanna, *Flexible Electronics, Volume 2 – Thin-film transistors*. 2019.

[83]   R. V. Martinez, "Editorial for special issue on flexible electronics: Fabrication and ubiquitous integration," *Micromachines*, vol. 9, no. 11. MDPI AG, Nov. 19, 2018, doi: 10.3390/mi9110605

[84]   R. V. Martinez, *Flexible Electronics. Fabrication and Ubiquitous Integration*. 2019

[85]   J. C. Agar, K. J. Lin, R. Zhang, J. Durden, K. S. Moon, and C. P. Wong, "Novel PDMS(silicone)-in-PDMS(silicone): Low cost flexible electronics without metallization," 2010, doi: 10.1109/ECTC.2010.5490654

[86]   H. Wu, C. Yang, J. Liu, X. Cui, B. Xie, and Z. Zhang, "A highly conductive thermoplastic electrically conductive adhesive for flexible and low cost electronics," 2014, doi: 10.1109/ICEPT.2014.6922948

[87]   F. F. Vidor, T. Meyers, U. Hilleringmann, and G. I. Wirth, "Influence of UV irradiation and humidity on a low-cost ZnO nanoparticle TFT for flexible electronics," 2015, doi: 10.1109/NANO.2015.7388836

[88]   J. Reker, T. Meyers, F. F. Vidor, T. H. Joubert, and U. Hilleringmann, "Integration Process for Self-Aligned Sub-μm Thin-Film Transistors for Flexible Electronics," 2021, doi: 10.1109/FLEPS51544.2021.9469764

[89]   N. S. Korivi and J. Li, "Metal patterning on polymers for flexible microsystems and large-area electronics," 2007, doi: 10.1109/SSST.2007.352344

[90]   M. Soni, D. Shakthivel, A. Christou, A. Zumeit, N. Yogeswaran, and R. Dahiya, "High Performance Printed Electronics on Large Area Flexible Substrates," 2020, doi: 10.1109/EDTM47692.2020.9118012

[91]   T. Someya, T. Sakurai, T. Sekitani, and Y. Noguchi, "Printed organic transistors for large-area electronics," 2007, doi: 10.1109/POLYTR.2007.4339127

[92]   L. G. Occhipinti, "Innovative manufacturing of large-area electronics," 2014, doi: 10.1109/ESSDERC.2014.6948793

[93]   C. M. Wang, B. Hou, Y. K. Deng, M. B. Zhou, and X. P. Zhang, "Large-area Die Attachment and the Surface Finish Effect on Bonding Strength of Joints in High-power Electronics Using a Low-temperature Sinterable Cu Nanoparticle Paste," 2022, doi: 10.1109/ICEPT56209.2022.9873390

[94]   L. Roselli *et al.*, "Smart surfaces: Large area electronics systems for internet of things enabled by energy harvesting," *Proc. IEEE*, vol. 102, no. 11, 2014, doi: 10.1109/JPROC.2014.2357493

[95]   K. L. Lin and K. Jain, "Design and fabrication of stretchable multilayer self-aligned interconnects for flexible electronics and large-area sensor arrays using excimer laser photoablation," *IEEE Electron Device Lett.*, vol. 30, no. 1, 2009, doi: 10.1109/LED.2008.2008665

[96] B. Salam, X. C. Shan, and W. Jun, "Large area roll-to-roll screen printing of electrically conductive circuitries," 2017, doi: 10.1109/EPTC.2016.7861481

[97] N. Verma *et al.*, "Enabling scalable hybrid systems: Architectures for exploiting large-area electronics in applications," *Proc. IEEE*, vol. 103, no. 4, 2015, doi: 10.1109/JPROC.2015.2399476

[98] F. Y. Kuo, C. Y. Lin, P. C. Chuang, C. L. Chien, Y. L. Yeh, and S. K. A. Wen, "Monolithic Multi-Sensor Design with Resonator-Based MEMS Structures," *IEEE J. Electron Devices Soc.*, vol. 5, no. 3, 2017, doi: 10.1109/JEDS.2017.2666821

[99] M. Hautefeuille, B. O'Flynn, F. Peters, and C. O'Mahony, "Miniaturised multi-MEMS sensor development," *Microelectron. Reliab.*, vol. 49, no. 6, 2009, doi: 10.1016/j.microrel.2009.02.017

[100] N. Banerjee, A. Banerjee, N. Hasan, S. S. Pandey, B. P. Gogoi, and C. H. Mastrangelo, "A monolithically integrated multi-sensor platform," 2015, doi: 10.1109/ICSENS.2015.7370324

[101] M. Mansoor, I. Haneef, S. Akhtar, M. A. Rafiq, S. Z. Ali, and F. Udrea, "SOI CMOS multi-sensors MEMS chip for aerospace applications," in *Proceedings of IEEE Sensors*, 2014, vol. 2014-December, no. December, doi: 10.1109/ICSENS.2014.6985225

[102] A. Hyldgård, K. Birkelund, J. Janting, and E. V. Thomsen, "Direct media exposure of MEMS multi-sensor systems using a potted-tube packaging concept," *Sensors Actuators, A Phys.*, vol. 142, no. 1, 2008, doi: 10.1016/j.sna.2007.02.024

[103] Q. Ma, Z. Wang, and L. Pan, "Monolithic integration of multiple sensors on a single silicon chip," 2016, doi: 10.1109/DTIP.2016.7514831

[104] J. M. Nassar, G. A. T. Sevilla, S. J. Velling, M. D. Cordero, and M. M. Hussain, "A CMOS-compatible large-scale monolithic integration of heterogeneous multi-sensors on flexible silicon for IoT applications," 2017, doi: 10.1109/IEDM.2016.7838448

[105] A. Cardoso, S. Kroehnert, R. Pinto, E. Fernandes, and I. Barros, "Integration of MEMS/Sensors in Fan-Out wafer-level packaging technology based system-in-package (WLSiP)," 2017, doi: 10.1109/EPTC.2016.7861591

[106] K. W. Lee, "High-density fan-out technology for advanced SiP and 3D heterogeneous integration," in *IEEE International Reliability Physics Symposium Proceedings*, 2018, vol. 2018-March, doi: 10.1109/IRPS.2018.8353588

[107] M. F. Chen, F. C. Chen, W. C. Chiou, and D. C. H. Yu, "System on integrated chips (SoIC(TM) for 3D heterogeneous integration," in *Proceedings - Electronic Components and Technology Conference*, 2019, vol. 2019-May, doi: 10.1109/ECTC.2019.00095

[108]   S. W. Yoon, "Advanced 3D eWLB-SiP (embedded Wafer Level Ball Grid Array – System in Package) Technology," *Addit. Conf. (Device Packag. HiTEC, HiTEN, CICMT)*, vol. 2017, no. DPC, 2017, doi: 10.4071/2017dpc-tp2_presentation5

[109]   Y. Morikawa, "Advanced Package Wiring Technology Solution for Heterogeneous Integration," 2019, doi: 10.23919/PanPacific. 2019.8696855

[110]   P. Batude, T. Ernst, J. Arcamone, G. Arndt, P. Coudrain, and P. E. Gaillardon, "3-D sequential integration: A key enabling technology for heterogeneous Co-integration of new function with CMOS," *IEEE J. Emerg. Sel. Top. Circuits Syst.*, vol. 2, no. 4, 2012, doi: 10.1109/JETCAS.2012.2223593

[111]   B. Debnath and R. Kumar, "A Comparative Simulation Study of the Different Variations of PZT Piezoelectric Material by Using A MEMS Vibration Energy Harvester," *IEEE Trans. Ind. Appl.*, vol. 58, no. 3, 2022, doi: 10.1109/TIA.2022.3160144

[112]   L. H. Fang, S. I. S. Hassan, R. B. A. Rahim, and J. M. Nordin, "A review of techniques design acoustic energy harvesting," 2015, doi: 10.1109/SCORED.2015.7449358

[113]   J. Iannacci, G. Sordo, M. Schneider, U. Schmid, A. Camarda, and A. Romani, "A novel toggle-type MEMS vibration energy harvester for internet of Things applications," 2016, doi: 10.1109/ICSENS.2016.7808553

[114]   D. Galayko, A. Dudka, and P. Basset, "Capacitive kinetic energy harvesting: System-level engineering challenges," 2014, doi: 10.1109/ICECS.2014.7050127

[115]   S. Niu, C. Gao, and J. Chen, "Modeling and Analysis of a Non-linear Electret Electrostatic Energy Harvester," 2020, doi: 10.1109/AUTEEE50969.2020.9315731

[116]   Y. Li, J. Li, A. Yang, Y. Zhang, B. Jiang, and D. Qiao, "Electromagnetic Vibrational Energy Harvester with Microfabricated Springs and Flexible Coils," *IEEE Trans. Ind. Electron.*, vol. 68, no. 3, 2021, doi: 10.1109/TIE.2020.2973911

[117]   D. Han, M. Kine, T. Shinshi, and S. Kadota, "MEMS Energy Harvester Utilizing a Multi-pole Magnet and a High-aspect-ratio Array Coil for Low Frequency Vibrations," 2020, doi: 10.1109/powermems49317.2019.51289501055

[118]   K. Wang, G. Wang, X. Dai, G. Ding, and X. Zhao, "Implementation of Dual-Nonlinearity Mechanism for Bandwidth Extension of MEMS Multi-Modal Energy Harvester," *J. Microelectromechanical Syst.*, vol. 30, no. 1, 2021, doi: 10.1109/JMEMS.2020.3036901

[119] M. R. Awal, M. Jusoh, M. R. Kamarudin, T. Sabapathy, H. A. Rahim, and M. F. A. Malek, "Power harvesting using dual transformations of piezoelectricity and magnetism: A review," 2015, doi: 10.1109/SCORED.2015.7449392

[120] K. Wang, X. Dai, X. Xiang, G. DIng, and X. Zhao, "A MEMS-based Bi-Stable Electromagnetic Energy Harvester with an Integrated Magnetization-Reversible Circuit," 2019, doi: 10.1109/TRANSDUCERS.2019.8808386

[121] J. Cornett *et al.*, "Chip-scale thermal energy harvester using Bi2Te3," 2015, doi: 10.1109/IECON.2015.7392612

[122] E. T. Topal, H. Kulah, and A. Muhtaroglu, "Thin film thermoelectric energy harvesters for MEMS micropower generation," *2010 Int. Conf. Energy Aware Comput. ICEAC 2010*, pp. 2–5, 2010, doi: 10.1109/ICEAC.2010.5702321

[123] D. Rozgi and D. Markovi, "A 0.78mW/cm2 autonomous thermoelectric energy-harvester for biomedical sensors," in *IEEE Symposium on VLSI Circuits, Digest of Technical Papers*, 2015, vol. 2015-August, doi: 10.1109/VLSIC.2015.7231289

[124] A. A. Tahrim, A. Ahmad, M. Sultan, and M. Ali, "Silicon Nanowire Arrays Thermoelectric Power Harvester," pp. 728–731, 2017

[125] F. Islam, A. Zubair, and N. Fairuz, "Wearable Thermoelectric Nanogenerator Based on Carbon Nanotube for Energy Harvesting," 2019, doi: 10.1109/SCORED.2019.8896333

[126] V. Kotipalli *et al.*, "Carbon nanotube film-based cantilever for light and thermal energy harvesting," 2010, doi: 10.1109/ICSENS.2010.5690697

[127] M. Bobinger, S. Hinterleuthner, M. Becherer, S. Keddis, N. Schwesinger, and P. Lugli, "Energy harvesting from ambient light using PVDF with highly conductive and transparent silver nanowire/PEDOT:PSS hybride electrodes," 2017, doi: 10.1109/NANO.2017.8117272

[128] A. R. Ndjiongue and T. M. N. Ngatched, "LED-based energy harvesting systems for modern mobile terminals," 2020, doi: 10.1109/ISNCC49221.2020.9297232

[129] G. Moayeri Pour and W. D. Leon-Salas, "Solar energy harvesting with light emitting diodes," 2014, doi: 10.1109/ISCAS.2014.6865551

[130] M. S. Costa, L. T. Manera, and H. S. Moreira, "Study of the light energy harvesting capacity in indoor environments," 2019, doi: 10.1109/INSCIT.2019.8868516

[131] S. Kim *et al.*, "Ambient RF energy-harvesting technologies for self-sustainable standalone wireless sensor platforms," *Proc. IEEE*, vol. 102, no. 11, 2014, doi: 10.1109/JPROC.2014.2357031

[132] C. Merz, G. Kupris, and M. Niedernhuber, "Design and optimization of a radio frequency energy harvesting system for energizing low power

devices," in *International Conference on Applied Electronics*, 2014, vol. 2015-January, no. January, doi: 10.1109/AE.2014.7011703

[133] M. A. Bin Othman, "Waste of radio frequency signal analysis for wireless energy harvester," 2010, doi: 10.1109/CSPA.2010.5545241

[134] H. Ito *et al.*, "A 2.3 pJ/bit frequency-stable impulse OOK transmitter powered directly by an RF energy harvesting circuit with -19.5 dBm sensitivity," 2014, doi: 10.1109/RFIC.2014.6851645

[135] M. Aldrigo *et al.*, "Harvesting Electromagnetic Energy in the V-Band Using a Rectenna Formed by a Bow Tie Integrated with a 6-nm-Thick Au/HfO2/Pt Metal-Insulator-Metal Diode," *IEEE Trans. Electron Devices*, vol. 65, no. 7, 2018, doi: 10.1109/TED.2018.2835138

[136] F. Yu, J. Du, and X. Yang, "Four-Band Polarization-Insensitive and Wide-Angle Metasurface with Simplified Structure for Harvesting Electromagnetic Energy," 2019, doi: 10.1109/IEEE-IWS.2019.8804158

[137] F. Yu, X. Yang, and H. Zhong, "Polarization-Insensitive Wide-Angle-Reception Metasurface for Harvesting Electromagnetic Energy," 2019, doi: 10.23919/ACESS.2018.8669251

[138] N. Xing and G. A. Rincon-Mora, "Highest Wireless Power: Inductively Coupled or RF?," in *Proceedings - International Symposium on Quality Electronic Design, ISQED*, 2020, vol. 2020-March, doi: 10.1109/ISQED48828.2020.9136990

[139] X. Li, C. Y. Tsui, and W. H. Ki, "A 13.56 MHz Wireless Power Transfer System With Reconfigurable Resonant Regulating Rectifier and Wireless Power Control for Implantable Medical Devices," *IEEE J. Solid-State Circuits*, vol. 50, no. 4, 2015, doi: 10.1109/JSSC.2014.2387832

[140] S. Gautam, S. Kumar, S. Chatzinotas, and B. Ottersten, "Experimental Evaluation of RF Waveform Designs for Wireless Power Transfer Using Software Defined Radio," *IEEE Access*, vol. 9, 2021, doi: 10.1109/ACCESS.2021.3115048

[141] M. Kim, H. S. Lee, and H. M. Lee, "An Efficient Wireless Power and Data Transfer System with Current-Modulated Energy-Reuse Back Telemetry and Energy-Adaptive Dual-Input Voltage Regulation," in *Proceedings of the Custom Integrated Circuits Conference*, 2021, vol. 2021-April, doi: 10.1109/CICC51472.2021.9431451

[142] V. Venkatesh, Q. Yang, J. Zhang, J. Pikul, and M. G. Allen, "FABRICATION AND CHARACTERIZATION OF EVAPORATED ZINC ANODES FOR SMALL-SCALE ZINC-AIR BATTERIES," in *21st International Conference on Solid-State Sensors, Actuators and Microsystems, TRANSDUCERS 2021*, 2021, vol. 2021-January, doi: 10.1109/TRANSDUCERS50396.2021.9495470

[143] A. Kornyushchenko, V. Natalich, S. Shevchenko, and V. Perekrestov, "Formation of Zn/ZnO and Zn/ZnO/NiO Multilayer Porous Nanosystems

for Potential Application as Electrodes in Li-ion Batteries," 2020, doi: 10.1109/NAP51477.2020.9309614

[144]   G. Ouyang, G. Whang, E. MacInnis, and S. S. Iyer, "Fabrication of Flexible Ionic-Liquid Thin Film Battery Matrix on FlexTrateTM for Powering Wearable Devices," in *Proceedings - Electronic Components and Technology Conference*, 2021, vol. 2021-June, doi: 10.1109/ECTC32696.2021.00257

[145]   M. Badr, M. M. Aboudina, F. A. Hussien, and A. N. Mohieldin, "Simultaneous Multi-Source Integrated Energy Harvesting System for IoE Applications," in *Midwest Symposium on Circuits and Systems*, 2019, vol. 2019-August, doi: 10.1109/MWSCAS.2019.8884893

[146]   X. Cui, J. Zhang, H. Zhou, and C. Deng, "PowerPool: Multi-source Ambient Energy harvesting," 2020, doi: 10.1109/BigCom51056.2020.00019

[147]   W. Zhou *et al.*, "Research on Multi-source Environmental Micro Energy Harvesting and Utilization," 2021, doi: 10.1109/ACPEE51499.2021.9437121

[148]   J. Li, J. Hoon Hyun, and D. Sam Ha, "A multi-source energy harvesting system to power microcontrollers for cryptography," 2018, doi: 10.1109/IECON.2018.8591833

[149]   V. Stomelli *et al.*, "A Multi-Source Energy Harvesting Sensory Glove Electronic Architecture," 2018

[150]   J. Yan, X. Liao, S. Ji, and S. Zhang, "A Novel Multi-Source Micro Power Generator for Harvesting Thermal and Optical Energy," *IEEE Electron Device Lett.*, vol. 40, no. 2, 2019, doi: 10.1109/LED.2018.2889300

[151]   Y. Zhu, D. F. Wang, Y. Fu, and X. Yang, "Developing self-powered high performance sensors: Part i - A tri-axial piezoelectric accelerometer with an error compensation method," 2018, doi: 10.1109/DTIP.2018.8394223

[152]   L. Zheng, D. F. Wang, Y. Fu, X. Yang, Y. Suzuki and M. Ryutaro, "Developing Self-powered High Performance Sensors: Part II - Preliminary Study On PE-ME Coupling In A Vibration Energy Convertor," 2020, doi: 10.1109/DTIP51112.2020.9139142

[153]   C. Li, Q. Lan, D. F. Wang, T. Itoh, and R. Maeda, "Developing Self-powered High Performance Sensors: Part III-A Low Frequency Ball-impacted PE/ME Composite Energy Harvester," 2022, doi: 10.1109/DTIP56576.2022.9911741

[154]   R. Frei, R. McWilliam, B. Derrick, A. Purvis, A. Tiwari, and G. Di Marzo Serugendo, "Self-healing and self-repairing technologies," *Int. J. Adv. Manuf. Technol.*, vol. 69, no. 5–8, 2013, doi: 10.1007/s00170-013-5070-2

[155]   C. Lesaint *et al.*, "Self-healing high voltage electrical insulation materials," 2014, doi: 10.1109/EIC.2014.6869384

[156] X. Wang, Z. Chen, W. Xu, and X. Wang, "Fluorescence labelling and self-healing microcapsules for detection and repair of surface microcracks in cement matrix," *Compos. Part B Eng.*, vol. 184, 2020, doi: 10.1016/j.compositesb.2020.107744

[157] N. Xu et al., "Employing ultrasonic wave as a novel trigger of microcapsule self-healing cementitious materials," *Cem. Concr. Compos.*, vol. 118, 2021, doi: 10.1016/j.cemconcomp.2021.103951

[158] M. W. Lee, S. An, S. S. Yoon, and A. L. Yarin, "Advances in self-healing materials based on vascular networks with mechanical self-repair characteristics," *Advances in Colloid and Interface Science*, vol. 252. 2018, doi: 10.1016/j.cis.2017.12.010

[159] T. Selvarajoo, R. E. Davies, B. L. Freeman, and A. D. Jefferson, "Mechanical response of a vascular self-healing cementitious material system under varying loading conditions," *Constr. Build. Mater.*, vol. 254, 2020, doi: 10.1016/j.conbuildmat.2020.119245

[160] R. Luterbacher, T. S. Coope, R. S. Trask, and I. P. Bond, "Vascular self-healing within carbon fibre reinforced polymer stringer run-out configurations," *Compos. Sci. Technol.*, vol. 136, 2016, doi: 10.1016/j.compscitech.2016.10.007

[161] L. Q. Zhou et al., "Ultrasonic Inspection and Self-Healing of Ge and 3C-SiC Semiconductor Membranes," *J. Microelectromechanical Syst.*, 2020, doi: 10.1109/JMEMS.2020.2981909

[162] Y. Zhang, H. Khanbareh, J. Roscow, M. Pan, C. Bowen, and C. Wan, "Self-Healing of Materials under High Electrical Stress," *Matter*, vol. 3, no. 4. 2020, doi: 10.1016/j.matt.2020.07.020

[163] V. Kumar, X. Guo, R. Jafari, and S. Pourkamali, "Ultra-low power self-computing binary output digital MEMS accelerometer," in *Proceedings of the IEEE International Conference on Micro Electro Mechanical Systems (MEMS)*, 2016, vol. 2016-February, doi: 10.1109/MEMSYS.2016.7421607

[164] J. Iannacci, "Towards Future 6G from the Hardware Components Perspective - A Focus on the Hardware-Software Divide, its Limiting Factors and the Envisioned Benefits in Going beyond it," 2021, doi: 10.1109/5GWF52925.2021.00008

[165] J. Iannacci, "A Perspective Vision of Micro/Nano Systems and Technologies as Enablers of 6G, Super-IoT, and Tactile Internet," *Proc. IEEE*, vol. 111, no. 1, 2023, doi: 10.1109/JPROC.2022.3223791

CHAPTER

3

A Conceptual Framework to Conceive 6G/FN-enabling Micro-devices and Nano-devices

Abstract

This conclusive chapter is going to address the actual reformulation of the concept of HW, along with its formalization. First, the concept of the *HW–SW divide* will be introduced, highlighting the cornerstone approaches to the development of HW–SW systems, which have remained unaltered for about five decades of telecommunications, and that can turn into a limiting factor when facing the challenges ahead of 6G/FN. Subsequently, the concepts of *separation* and *symmetry*, and in particular their increase, will be discussed and reported as a solution to overcome the limitations of the HW–SW divide. Starting from such a base of knowledge, a conceptual framework for the development of future micro-/nano-devices with unprecedented characteristics (as demanded from 6G/FN) will be introduced. It is named the WEAF Mnecosystem (water, earth, air and fire micro/nanotechnologies ecosystem). Such a tool capitalizes upon an analogy between the features of low-complexity HW components and the four elements in nature.

3.1 The Hardware–Software Divide and What is Beyond

Since the uptake of semiconductor-based devices and technologies, the fields of electronics, telecommunications and computer science have known relentless expansion and evolution, which have been going on for about six decades now. As already pointed out, silicon-based HW technologies have progressed

75

according to Moore's law [1–4], which has remarkably described the relationship between the size scaling down of HW devices, due to the evolution of micro-fabrication technologies/processes, and the increase of computational capability per unit area. On a different plane of reference, HW technologies also developed according to paths transcending the mainstream direction of silicon devices scaling down. Those alternative developments fall under the classification of more than Moore technologies [5–8], and address, for instance, micro/nanosystems and micro/nanotechnologies (MEMS/NEMS) [9–12], novel materials [13, 14], heterostructure-based transistors [15], nanoelectronics [16], as well as packaging [17–20], 3D stacking and integration solutions and methodologies [21–23].

In a few brief words, the whole development of electronic HW technologies has been led and dominated for several decades by what the proposition *more Moore and more than Moore* effectively frames [24].

3.1.1 Defining the hardware–software divide

The combination of more Moore and more than Moore drivers has been leading fundamental trends in electronics, like miniaturization, integration, reconfigurability and power consumption decrease. Yet, within this sort of seamless revolution which has been relentlessly advancing for decades, there exists a cornerstone item that has remained unchanged since the very beginning, but still holds validity now. Given that each electronic system, sub-system or building block is a blend of HW and SW, a radical asymmetry lays in their definition.

The SW, due to its immateriality, is something that can be easily fixed, upgraded or replaced, (virtually) at any time, when the system has been already deployed, often without interrupting its normal (at least basic) operation. Differently, the HW is hard-bound to its physical dimension. Fixing or upgrading the HW means, most of times, replacing or adding something to the system. There is no question that HW can be reconfigured, especially in modern emerging applications. Also, it can redundantly ensure operation, e.g., in safety/mission critical contexts. Nonetheless, to do so, the HW has to be, respectively, reconfigurable and redundant. In other words, conversely to SW, the flexibility/adaptivity of HW operation has somehow to be accounted for at the design and deployment level, certainly not at the system operation level (apart from unexpected failure conditions).

Bearing these concepts in mind, sub-systems and systems are designed according to the HW–SW co-design philosophy [25–28]. This means that the physical device, for instance a radio transceiver, and the algorithms, are jointly conceived, aiming at optimal operation of the overall system. In doing so, the HW is tailored around the functionalities and characteristics offered by the SW, leading to the most reasonable compromises in terms of performance, compliance of specifications, reliability, complexity and cost of the final product.

The asymmetry just described is what is termed in this work as the *HW–SW divide*. Based on this principle, telecommunication standards evolved, across several decades, from 1G to 5G, without any substantial need for its revision, modification or extension.

Nonetheless, as mentioned before, the 5G is already today expected not to fully meet the challenge of being the factual enabler of IoE and super-IoT [29]. This work speculates that a likely reason for such an early downsizing of expectations is the persistence of the HW–SW divide philosophy in the current development of 5G services and physical infrastructure.

3.1.2 Looking at 6G and FN – beyond the hardware–software divide

The purpose of this section is that of formulating the *question* emerging from what was discussed up to here. That said, the WEAF Mnecosystem, to be discussed in the following pages, will attempt to provide a comprehensive answer.

The way the currently under deployment 5G addresses flexibility and adaptivity is through virtualization. To this end, two pillar technologies are SDN [30–33] and NFV [34–36], pointing in the direction of SW-based virtual networks that, while relying upon physical (HW) infrastructures, gain significant features in terms of reconfigurability and wide usability, among which is network slicing [37–39].

The crucial point under observation here is that the mentioned *softwarization* chased by 5G is lacking and inappropriate to address the network *intelligentization* that will be embraced by 6G through incorporation of AI [40]. As a matter of fact, 6G will have to support diverse, heterogeneous and often orthogonal functions, like, e.g., communications (from GHz, to mm-waves, to THz), computing, energy distribution (via EH and WPT), as well as sensing, data collection and storage, analytics and computation.

Such a scenario, in its entirety, reveals the intrinsic limitations imposed by the HW–SW divide, also including the HW–SW co-design approach. At the same time, the need for a modification of the classical set of conceptual and factual interlinks between the HW and SW entities, emerges rather clearly. This kind of reformulation should pass through the enhancement of the concepts of *separation* and *symmetry*, described as follows:

- **Separation**. In contrast to the HW–SW co-design approach, in which both HW and SW are optimized around a certain set of functionalities and specifications, the algorithm–hardware separation trend will have to be pursued and significantly pushed forward. This is a pivotal approach to address the heterogeneous, upgradable and adaptable HW capabilities necessary to blend together all the diverse services mentioned before, as well as to support, on the other hand, the network self-sustaining philosophy boosted by AI.
- **Symmetry**. Within the landscape of an increasing HW–SW separation, more symmetry between these two entities will also be of paramount importance. This means that, differently from the current conception, the HW will have to acquire more flexibility, self-adaptivity and intelligence, thus thinning the gap to SW. In other words, 6G and FN will demand an increased level of abstraction of the HW, aiming to weaken its currently tight constraints to the physical dimension.

A visual representation of the concepts of separation and symmetry, concerned with the HW–SW set of relationships, is proposed in the following Figure 3.1 and Figure 3.2, respectively.

Stepping into more detail, Figure 3.1 (a) shows the conceptual scheme of a typical system conceived relying upon the classical HW–SW co-design concept. Diversely, Figure 3.1 (b) sketches what it is envisaged to happen in future with 6G and FN if leveraging increased separation between HW and SW. Therefore, the content of Figure 3.1 can be summarized as follows. Figure 3.1 (a) is the case of one HW module and one SW module, optimized with respect to each other to realize specific functions defined ab-initio. On the other hand, Figure 3.1 (b) sketches the case of many HW and many SW modules that combine together dynamically to implement a variety of different and not a-priori-defined functions. In addition, Figure 3.2 focuses on the concept of symmetry between HW and SW.

In detail, Figure 3.2 (a) sketches the current situation, typical of the HW–SW co-design approach, in which an upper boundary is ideally set to the *intelligence* that HW can achieve, as intrinsically originating by its definition. The term intelligence is used with a rather broad meaning here, as it addresses features like flexibility, reconfigurability, replaceability, along with self-referred characteristics, like self-repairability, self-adaptability, etc. That said, current systems are based on a marked asymmetry between HW and SW intelligence, with a ceiling that cannot be intrinsically surpassed by the HW, and above which lays the SW.

Figure 3.1: (a) Scheme of systems based on the classical HW–SW co-design approach, in which HW and SW verticals are optimized with respect to each other. (b) Scheme of what HW–SW systems will be if leveraging more separation between them, with diverse HW and SW modules for a plethora of diversified and dynamic systems. (Images taken from [41] and published under a Creative Commons Attribution 4.0 License.)

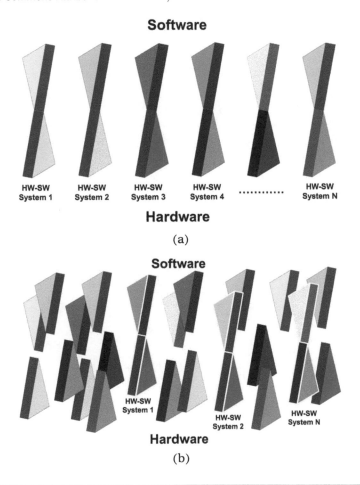

The novel scenario envisaged in this work is the one depicted in Figure 3.2 (b), where the boundary to intelligence is weakened, and possibly removed, letting the HW raise to higher levels, as well as the SW to move lower, if necessary in any circumstance.

As a partial conclusion of this section and of the previous pages, the scenario sketched above is at the moment a vision of what 6G and, most of all, on a longer

Figure 3.2: Schematic of the concept of symmetry between HW and SW. (a) Standard approaches, like the HW–SW co-design, define a ceiling (upper boundary) to the level of intelligence reached by the HW, while whatever is above such a plane is the floor of the SW domain. (b) Future envisaged scenario, in which the mentioned boundary is weakened, and eventually removed. (Images taken from [41] and published under a Creative Commons Attribution 4.0 License.)

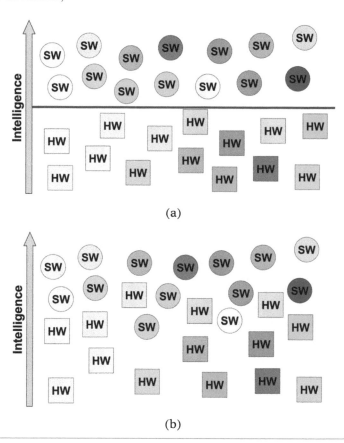

(a)

(b)

term basis, FN could possibly be. Yet, it is not unliked from what is discussed in literature. To this end, a few relevant considerations are reported in [40] for what concerns RF transceivers. It is stated that HW capabilities, like number of antennas, RF chains, the resolution and sampling rates of ADC, etc., remained quasi-static in the jointly designed HW–SW systems, from 1G to the current 5G. On the other hand, the work highlights that recent state-of-the-art research in circuits and antennas is relevantly improving the HW capabilities, with the target of enabling 6G BS and handsets to be diversified and upgradable.

Finally, the following sentence in [40] is of particular relevance in light of the scope of this work: "6G will not be operating under the conventional joint design, which fails in allowing agile adaptation to a diversified and upgradable hardware".

3.2 Introducing the WEAF Mnecosystem

One of the main targets of this work is that of addressing the emerging mismatch between the HW–SW divide and what the AI-driven 6G, super-IoT and TI will be demanding. As suggested in the previous section, increased separation and symmetry of HW and SW will have to be purposed.

It is argued here that the way such trends can be pursued is through the reformulation of the concept of hardware (HW), according to which, despite still based on its physical (tangible) dimensions, it acquires a higher level of abstraction, thus thinning the existing gap with the SW.

As reported in more technical detail later, the HW reconceptualization capitalizes on both consolidated and emerging microtechnologies and nanotechnologies, encompassing micro/nanomaterials, micro/nanosystems and micro/nanoelectronics, identified as KET bearing disruptive potential to make the transition to 6G turn into reality.

Driven by the twofold aim of simplifying the explanation and building a comprehensive frame of reference, analogy between HW–SW systems and the four classical elements [42],[43] is proposed by this work. The envisioned ecosystem is named the WEAF Mnecosystem, standing for the water, earth, air and fire micro/nanotechnologies ecosystem, visually represented in Figure 3.3.

Within the WEAF Mnecosystem scenario, earth and air refer to the classical concepts of HW and SW, respectively. Differently, water and fire transcend the common meaning of HW. In particular, water indicates *transmorphic* HW able to self-adapt, reconfigure and evolve, thereafter gaining some features typical of the SW. On the other hand, fire embraces the concept of energy needed to sustain operation of parts of the system, with particular reference to the edge of the network. The HW supporting energy has to ensure its storage, conversion and transportation, with proper amount and timeliness, i.e., where needed, when needed.

In the following subsections, deeper insight into the WEAF elements conception tailored around the HW is going to be developed. Since earth and air represent the classical concepts of HW and SW, respectively, their description is going to be rather brief. By contrast, more room will be reserved for water and fire, as they bear most of innovation content.

Figure 3.3: Symbolic description of the WEAF Mnecosystem, based on the four classical elements: water, earth, air and fire.

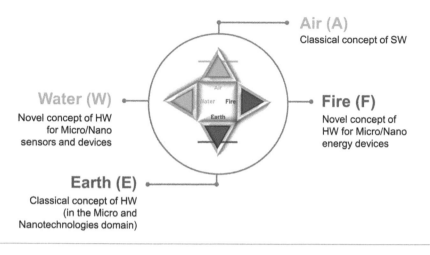

3.2.1 Earth and air

As earth and air are the standard concepts of HW and SW, respectively, their combination yields the consolidated approach of HW–SW co-design already mentioned in this work. Thereafter, there are no significant aspects to be discussed in addition to what current design strategies and state-of-the-art are addressing.

From a different perspective, instead, it is important to highlight that earth and air are parts of the WEAF Mnecosystem, as well as water and fire. Belonging to the same ecosystem, classical HW, SW, as well as co-designed HW–SW systems, are free to interact with other elements, as is going to be discussed in the following. In fact, within the frame of the WEAF Mnecosystem, earth and air are pivotal elements, nurturing the conception of novel water and fire HW items.

3.2.2 Water

Starting from a general point of view, water addresses pieces of HW that, similarly to the classical natural element, are physical and touchable, yet liquid and not provided with a particular shape, which equals the characteristic of gaining different shapes (i.e., any, in the case of water).

Entering now a more practical scenario, the water-based HW reconceptualization builds upon five fundamental concepts. The first three are intrinsic to the novel concept of HW itself, and are named: differentiation, evolution and ubiquity. The remaining two, instead, stem from osmotic interactions of HW and SW. Because of such an interplay, they are addressed as phase changes, namely, evaporation and condensation. A more comprehensive discussion around the just mentioned five constitutive concepts of water is now developed.

3.2.2.1 Differentiation

This feature addresses enhanced functional adaptivity, and refers to the ability of a single basic HW component to implement multiple orthogonal functions. A practical reference example could be a miniaturized actuator, like a MEMS micro-relay, also able to operate as an inertial sensor, as well as an EH, converting part of the environmental vibrations scattered energy into electricity. Commuting from one to another function of the HW component will be driven by the contingency needs that might emerge in the local spot of the 6G infrastructure, where that MEMS multi-functional device is deployed.

3.2.2.2 Evolution

Similarly to the previous item, this feature adds *liquidity* to the HW. However, this time the water-like acquired characteristic is not the functional differentiation, despite the HW capacity of evolving in physical terms. As it is easy to figure out, HW cannot grow or self-generate out of nowhere. Nonetheless, leveraging redundancy, as well as ease of integration and inexpensiveness typical of microtechnologies and nanotechnologies, it could be possible to surround each micro-/nano-sensor/actuator with a variety of other ghost devices. Such redundant items are latent (and basically useless) with respect to nominal functionality. In case new ways of operation emerge in a certain spot at the edge of the 6G network, those latent HW components can be waken up and wisely connected together in order to implement new and not predicted ab-initio functionalities, i.e., in turn, services. Still capitalizing on the principle of redundancy, such a self-evolutionary capacity encompasses self-repairing features [44] as well. A zero-level unsophisticated strategy to score this goal would be to simply rely on switching from a malfunctioning device or sub-system to a fresh (redundant) one.

3.2.2.3 Ubiquity

At a higher level of abstraction, the synergic combination of differentiation and evolution enables the ability of water of gaining different shapes and of flowing anywhere in the surrounding space, i.e., of being ubiquitous. In the case of water HW, physical devices will not be of course traveling from one spot to another of the infrastructure. Nevertheless, leveraging their functional differentiation and redundancy, of both identical and diverse components, a certain sensing or actuating feature could be functionally *moved* freely here and there at operation level, thus triggering (factual) ubiquity.

3.2.2.4 Evaporation

The first phase change can also be addressed as HW–SW (water–air) upstream. Such a feature starts sketching a close and mutual interaction existing between HW and SW, according to the analogy in nature between water and air. In particular, in this case the phase transition of water into air is at stake. Within such a frame, HW can gain some intelligence to self-adapt and self-optimize against time-changing conditions, instead of being controlled and reconfigured by ad-hoc SW algorithms running on dedicated control circuitry. The HW–SW upstream trend would lead to significantly simplified network edge ramifications and, in turn, to improved agility of operation and adaptivity.

A practical example can be that of HW units managing the switching across RF channels. Current solutions rely on the presence of redundant relays, controlled by SW routines and activated in case of need. Conversely, solutions based upon the water–air upstream (evaporation) could employ micro-switches able to self-adapt to the RF power levels, getting rid of dedicated control circuitry and embedded intelligence. This could be achieved capitalizing on the physical properties of thin-films and of MEMS/NEMS, like thermal expansion and temperature-driven intrinsic mechanical stress relaxation.

3.2.2.5 Condensation

The second phase change can also be addressed as SW–HW (air–water) downstream. Such a feature complements evaporation and has to do with the phase transition of air into water. In practical terms, low-/medium-complexity SW routines and algorithms could be implemented in HW thanks to micro- and nanotechnologies [45]. As for the case of HW–SW upstream, condensation would bring more intelligence to the edge, as sketched by the main trends in 6G, while simplifying the overall architecture of the tiniest network ramifications.

3.2.3 Fire

Concerning the last element of the WEAF Mnecosystem, fire borrows the five main concepts just unrolled for water, i.e., differentiation, evolution and ubiquity, as well as both the phase changes (evaporation and condensation). The fundamental distinctive aspect is that in this case the focus is uniquely directed to energy. The backbone of the fire HW is that of raising the abstraction level of energy, making it flow when and where needed, as well as heat does in nature, transiting from warmer to cooler parts of a body.

This feature will be critical to sustain the operation of the network edge, which is expected to dramatically increase in terms of complexity and functionalities under 6G. As stated above, fire HW will widely capitalize on micro- and nano-structured devices for EH, WPT and storage, according to the same functional adaptivity and redundant philosophy discussed for water HW.

3.3 Targeted Review of WEAF Mnecosystem-related MEMS/NEMS Literature

Most parts of the discussion around the emerging approaches here at stake, were developed in the previous pages. The scope of the current conclusive section is to complete framing the pivotal role of microtechnologies and nanotechnologies in the scenarios of interest of 6G and FN. To this end, a brief recap of the state-of-the-art in MEMS/NEMS technologies is going to be constructed, according to the conceptual bases of the WEAF Mnecosystem.

3.3.1 Water-like micro- and nano-devices and solutions

Looking at the past and present literature, the already existing HW solutions that comply with one or more of the main characteristics of water-like devices, i.e., differentiation, evolution and ubiquity, are multi-functional sensors, actuators and transducers. As already pointed out before, this can be enabled both by a single HW component that implements orthogonal functions [46], as well as by multiple sensors, actuators and transducers. The latter context can be enabled in a twofold fashion. On one hand, diverse devices can be monolithically fabricated one beside another, within the same chip, by leveraging a unique micro-/nano-fabrication technology (SoC approach) [47],[48]. On the other hand,

HW devices realized in diverse and incompatible technologies can be packaged and integrated together, realizing the concept of SiP [49–51].

3.3.2 Fire-like micro-/nano-devices and solutions

The main characteristics of fire-like HW components are fully met by miniaturized devices for energy conversion, transfer and storage. These features have already been discussed in Section 2.4.4. Therefore, the main classes of devices implementing such functions are simply listed here for the sake of completeness.

Concerning conversion, miniaturized EH devices are of pivotal importance to address sources of energy available in the surrounding environment [41],[52]. In particular, mechanical vibrations, thermal gradients, light (both natural and artificial), along with electromagnetic radiation scattered in the environment, are effectively converted into electricity by micro/nano EH. Also importantly, energy can be deliberately transferred from one point to another, i.e., from where it is available and stored to where it is needed, by means of WPT solutions and systems. Closing the loop, storage of energy is also crucial. To this purpose, miniaturized thin-film highly-efficient batteries are crucial building blocks for the network edge of 6G and FN.

Eventually, as pointed out in Chapter 2, an enabling element of the mentioned diversified devices is that of realizing physical powering platforms, in which various EH, storage units, as well as with WPT systems, are brought together. Along this direction, putting together features of fire-like and water-like HW, the possibility of integrating different EH devices within the same SiP, as discussed in [53], is a solution that opens many directions for what concerns zero-power autonomous smart nodes at the edge.

3.3.3 Water–air interactions – evaporation and condensation

After recapping the intrinsic characteristics of water-like HW, the interactions between water and air are going to be briefly covered, according to the WEAF Mnecosystem conceptual frame of reference.

3.3.3.1 Evaporation

As has already emerged, the phase change of water–air upstream (evaporation), is effectively compiled by micro- and nano-devices exhibiting self-referred characteristics, i.e., functionalities realized by the physical device itself, without the need for a surrounding HW–SW system or sub-system. This is for instance the case of physical items with self-recovering capacities. Given such a framework, it must be highlighted that self-repairing is an inherently top-down process, i.e., linked to the need to replace a faulty component. On the other hand, self-healing is a bottom-up process, implying rehabilitation of a faulty component [44]. In addition, self-healing embodies, among other features, self-protection, self-inspection, self-management and self-awareness.

That said, self-repair appears as an intrinsic characteristic of HW–SW systems, which is well-known and widely exploited in modern applications. The step forth in the WEAF Mnecosystem is embedding self-healing properties within the micro- and nano-device multi-physical behavior, thus making the presence of a surrounding sub-system redundant and unnecessary.

Strategies and solutions in the literature, implementing self-restoration properties in the micro-/nano-world are numerous. Just to mention a very limited set of them, the work in [54] exploits CNTs fillers to trigger self-healing EMI shielding properties. Thanks to such a solution, isolation recovers from 16 dB to 30 dB after structural damage to the material is intentionally caused. In addition, the discussion in [55], already covered before in this work, leverages reservoirs with a self-healing liquid agent to recover from dielectric treeing due to high and prolonged electrostatic stress.

As also discussed in Chapter 2, thin-film semiconductor materials with self-healing properties are reported in [56]. In details, cubic SiC or Ge structures are able to self-reestablish mechanical resonance properties after the material is damaged. This is possible thanks to the formation of nano-bridges across separated regions, thus reducing voids due to cracking.

Relevant examples are also discussed in restoration of optical properties. To this end, the work reported in [57] demonstrates self-healing light guides for dynamic sensing (SHeaLDS). In particular, both intrinsic damage resilience of light propagation within the waveguide, and autonomously self-healing capacities of an elastomer are exploited. The structure is able to recover both from cutting and puncturing intentionally caused to the light propagating line.

Eventually, just to mention a different approach to self-healing, it is possible to act at design level as well. To this end, the work in [58] discusses a microsystems-based vibration EH (EH-MEMS). In order to counteract the damage that typically occurs where the mechanical strain is more intense, the piezoelectric layer for energy conversion is arranged as an array, rather than being continuous. This way, smaller capacitors are deployed where the mechanical strain is larger, i.e., where damage is more likely to occur. In the case of cracks across shorted elements, failed capacitors can be excluded according to the anti-fuse approach, thus ensuring operation of the EH. At 1 kHz vibrations and 1 kΩ load, the EH-MEMS generates 100 mV. Excluding the damaged capacitors after failure, the output voltage drops to 87 mV, yet the device still works.

3.3.3.2 Condensation

Focusing on the phase change of air–water downstream (condensation), the examples in the literature concerned with items complying with its characteristics are very numerous. They can be reconducted to a couple of main classes of devices, those being micro/nano logic gates and memory cells. Therefore, also recalling the previous discussion in Chapter 2, it is unnecessary to report diverse works. By contrast, for the benefit of briefness, a unique reference example per each class of devices is going to be reviewed.

Concerning miniaturized logic gates, the work in [59] reports a clamped-clamped electrostatically actuated MEMS bar (600 µm long). Multiple electrodes are deployed around the resonating structure for the application of logic inputs, thus increasing the available DoF. The output information is carried by the activation/deactivation of the second mode of vibration. The time per operation is around 35 ms, while the power consumption is around 10–13 fJ.

Finally, when micro-/nano-structured memory cells are concerned, a relevant example is discussed in [60]. The mechanical transducers at stake are CMOS compatible NEMS cantilevers and clamped-clamped membranes. Electrostatic nano-switches store information in the ON/OFF input/output state.

The inherent limitation of electrostatically operated micro-mechanical switches is that the ON/OFF hysteretic behavior is monostable. This means that the device holds stable in its rest position (OFF – "0" logic state). Differently, when actuated (ON – "1" logic state), the DC controlling voltage must always be maintained, otherwise the relay commutes back to the OFF state. In other words, the OFF state is stable, while the ON state is unstable. The implication of monostability is that the memory cell consumes energy when kept in the ON state.

The solution in [60] solves this problem quite effectively. In fact, proper amount of electric charge is injected in a thin-film insulator beneath the NEMS membranes. This builds an intrinsic potential, which screens the applied DC bias. The result is that the pull-in/pull-out characteristics is shifted, making the memory cells bistable. This means that the memory cell requires energy only for commuting from "0" to "1" or vice-versa, but no longer for holding the "1" logic information.

3.3.4 Review of earth-like micro- and nanotechnologies

In the previous discussion of the WEAF Mnecosystem conceptual framework, it was highlighted that the earth element refers to the standard conception of HW, while water and fire embody mostly the innovation related to unprecedented micro- and nano-devices for 6G and FN. This statement is not equal to saying that earth-like devices play an ancillary role within the ecosystem, but rather the opposite. In fact, earth technologies are the playground from which it is possible to pick solutions that, properly extended and enriched of functionalities, lead to novel water-like and fire-like micro- and nano-devices.

In light of the above preamble, a few relevant examples of earth-like technologies is going to be provided below. Of course, the following content must not be intended as a comprehensive list of entries, but a preliminary and partial description of items carrying significant potential to sustain the WEAF Mnecosystem.

3.3.4.1 MEMS-based radio frequency passives (RF-MEMS)

As seen in Chapter 1, the exploitation of MEMS technology for the realization of RF passive components with pronounced characteristics in terms of high-performance and wide reconfigurability was covered. That said, RF-MEMS technology has been exploited for a couple of decades to realize miniaturized lumped components and complex networks. In the former category fall devices like micro-switches, both ohmic and capacitive, in series and shunt configuration, variable capacitors (varactors) and (tunable) inductors [61, 62].

Focusing on complex networks, a wide variety of multi-state reconfigurable and/or tunable highly miniaturized devices was realized and demonstrated in RF-MEMS technology. Among them, it is worth mentioning impedance matching tuners, filters, resonators, switching matrices, phase shifters (or delay lines), as well as step RF power attenuators and dividers [63, 64]. For all these classes of devices, remarkable characteristics in terms of large reconfigurability/tunability, frequency agility, low-loss, high-isolation, very-low power consumption, etc., were demonstrated [63]. Moreover, over the past several years, RF-MEMS technology commenced to be employed in mass-market applications [66], with 4G/5G smartphones first in line. In addition, MEMS-based RF passives have been recently characterized in the (sub-)THz range [67, 68]. The photograph in Figure 3.4 shows a silicon wafer of RF-MEMS devices manufactured in the technology platform available at Fondazione Bruno Kessler (FBK) in Italy [65].

Figure 3.4: Photograph of an entire silicon wafer populated by RF-MEMS devices manufactured at Fondazione Bruno Kessler (FBK) in Italy [65].

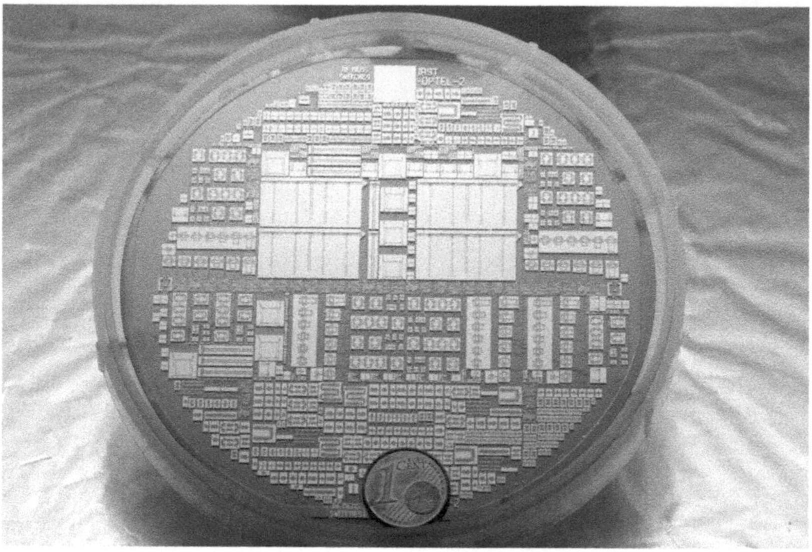

3.3.4.2 Metasurfaces and metamaterials

Metamaterials, also addressed with the term metasurfaces, realize physical characteristics, e.g., optical, acoustic and electromagnetic, of materials that typically do not exist in nature [69]. These features, united with the possibilities of miniaturization, are making metasurfaces particularly attractive for the realization of reconfigurable (smart) antennas [70], [71]. Just to mention a research example, the work in [72] discusses a planar lens antenna featuring a linear array of Tx/Rx elements for spatial beamforming and multibeam mMIMO. Lenses are formed by two-layer ultra-thin metamaterial surfaces, separated by air, and fed by a substrate-integrated waveguide-fed stacked patch antenna. An antenna scanning coverage of ±27° with gain tolerance of 3.7 dB is measured, with maximum gain of 24.2 dBi and aperture efficiency of 24.5 % in the 26.6–29 GHz frequency range.

3.3.4.3 Flexible electronics

The realization of miniaturized devices, like sensors, transducers, transistors, circuits and displays, on flexible substrates has been discussed in the literature for some time [73–75]. Flexible materials come both from thinning down of silicon wafers, or from employment of natively flexible substrates, like polymer foils. Moreover, technologies for flexible electronics enable low-cost and large-area (panel-like) manufacturing solutions. By virtue of these characteristics, devices and sub-systems on flexible substrates are gaining room in the field of IoT and wearable applications [76–78], while increasing chances of exploitation can be forecast for 6G and FN.

Focusing on a specific example in the literature, the work in [77] reports on the realization of thin-film transistors and electronics on flexible substrates by inkjet printing and spin-coating.

Silver interconnects of $100 \times 10\text{--}20 \ \mu m^2$ are realized, while the gate dielectric layer is obtained by vapor polymerization of Parylene. Optimizing ink formulation and thermal annealing of organic semiconductors, p-type characteristics with mobility above $1 \ cm^2/Vs$ are achieved.

3.3.5 Heterostructure-based semiconductors

Stepping into the broad field of more than Moore technologies (see Chapter 1), semiconductors that exploit hybridization of silicon with other materials must be mentioned. Such hetero-structured devices employ materials like graphene, Ge, In, SiC, GaAs, etc. [79–81]. Based on the used material/s, hetero-transistors achieve performance that is unknown to standard silicon-based (CMOS-like) technologies. Just to mention a few of them, boosted power handling [82] and higher electron mobility are scored. In particular, the latter one makes devices appropriate to work in high-frequency ranges, like mm-wave and above [83, 84]. Also important, hetero-structured semiconductor devices are open to low-cost processes [85], and, in turn, to flexible electronics techniques. In addition, vertical structuring of devices, rather than the classical planar approach, is becoming increasingly popular with this class of device. This eases the exploitation of novel deposition and patterning techniques, orthogonal with respect to those of classical CMOS processing. A relevant example, along such a direction, is that of FinFET transistors [86].

3.3.5.1 Device fusion through packaging and integration

The set of pros and cons linked to the realization of SoC, i.e., of sub-systems of various complexity, according to a monolithic manufacturing approach (unique technology), has already been covered in Chapter 2. In that context,

Figure 3.5: Generic image of an integrated electronic sub-system. (Image powered by Freepik.com.).

the exploitation of packaging and integration techniques (i.e., SiP) was also highlighted as a valid alternative to the SoC approach. That said, methods, techniques and technologies for encapsulation, shielding, protection and packaging of HW devices [87, 88], components and building blocks offer nearly unimaginable enabling solutions for integration of physical items within SiP [89], also capitalizing on 3D (vertical) stacking of modules [90]. In particular, the latter work discusses advanced WLP solutions for hetero-integrated wafer-level SiP and 3D heterogeneous integration. Such technologies enable integration of various functional dies, like, logics, memories, power IC, RF active chips and passive components. A generic visual representation of an integrated sub-system is shown in Figure 3.5.

3.3.5.2 Additive manufacturing (AM)

Placed orthogonally with respect to semiconductor technologies and micro-/nano-fabrication, the emerging fields of AM and 3D printing bear significant potential. Such terms address a heterogeneous technical ground of solutions and materials, like, e.g., powder or polymer based, electrically conductive and insulating, providing an extended set of DoF to developers and designers [91–95]. Given such a frame of reference, AM and 3D printing provide valuable potential in the 6G and FN scenarios. To this end, given the increasing trend to low-cost and large-surface manufacturing processes, hybridizations of AM

Figure 3.6: Generic image of a physical object realized by means of AM [Image powered by Freepik.com].

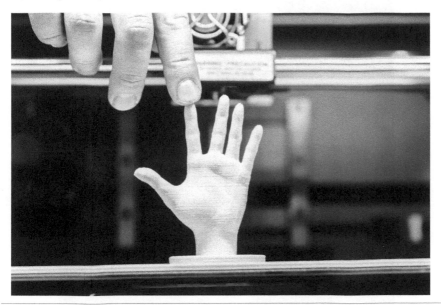

and 3D printing with one or more of the technologies covered before, are likely to be ventured in the future. Concerning practical exploitations, the work in [96] discusses AM technologies for the realization of RF passive components and antennas. 3D-printed microstrip lines with performances comparable to traditional copper tape (insertion loss below -5 dB up to 30 GHz) and microstrip antennas are reported. A photograph of a generic object manufactured through AM is shown in Figure 3.6.

3.3.5.3 Photonic devices

Looking into the scientific literature, the realization of optical devices by means of micro- and nanotechnologies has been successfully investigated for a long time. Items like micro-mirrors, (split) ring resonators, couplers, switches, often coupled to optical fibers [97–101], exploited as sensors, actuators and transducers, have been widely discussed and also exploited in market applications. Recalling 6G-driven KET, among which the exploitation of optical frequencies for communications, including VLC, must certainly be reported, a pivotal role can be envisaged for photonic devices.

Figure 3.7: Generic image of a prism. (Image powered by Freepik.com.)

That said, a MEMS-based micro-mirror obtained by bulk micromachining of SOI wafers with piezoelectric actuation is reported in [97]. The intrinsic mirror, torsion suspending structures and actuators are $400 \times 400 \ \mu m^2$, $20 \times 200 \ \mu m^2$, and $600 \times 1000 \ \mu m^2$, respectively, while the thickness of the entire device is 5.3 μm. The micro-mirror exhibits a 3.58° rotation when a 1 V peak-to-peak bias is applied at resonance (10.4 kHz). A generic image of an optical prism is shown In Figure 3.7.

3.3.5.4 Quantum technologies (QT)

Recently, the interest of research is getting more and more attracted by quantum phenomena and their exploitation in practical fields, like, e.g., sensing and computing. Also, QT match effectively with micro- and nanotechnology-based devices, aiming at practical applications in the fields of cryptography, telecommunications, computation, etc. Particular attention is being devoted to photonic devices working in the quantum domain, as well as to miniaturized objects merging optical and RF signals [102–105].

The work discussed in [106] shows a GaInN quantum well structure, exploited as opto-chemical transducer in (bio)chemical sensing. In contrast to conventional electrical read-out, purely optical photoluminescence read-out is performed. The scope of the sensor is to investigate the presence of an iron-storage ferritin complex with varying iron load, as it is generally considered as a superior biomarker for severe illnesses, such as Alzheimer's disease. GaInN quantum wells are promising candidates for next generation bioanalytical sensor structures.

3.3.6 Future Devices Inspired by the WEAF Mnecosystem – A Couple of Examples

Capitalizing on the articulate discussion developed across the previous pages, this conclusive section collects a couple of examples of functionally diversified devices that could be relevant in the application scenarios of 6G and FN. Such examples are inspired by the WEAF Mnecosystem conceptual framework.

3.3.7 RF-MEMS monolithic passive control units (PCU)

From the perspective of data transfer and communications, 6G will further leverage mMIMO antenna solutions, enabling advanced beamforming in the frame of small-, pico- and nano-cells [107] (see Figure 3.8).

The realization of high-order compact antenna arrays is proceeding at a good pace. However, this is just part of the whole picture. In fact, mMIMO need complex and redundant HW to be properly driven, both in Tx and Rx mode. To this end, TRM need to be interposed between the Tx/Rx mMIMO and the rest of the RF system, and must be physically duplicated for each of the antenna elements, expected to rise to 64 × 64 order and more. Moreover, operation at frequencies as high as sub-THz (100–300 GHz) will make the realization of high-performance TRM building blocks even more challenging, e.g., with respect to losses, power consumption, manufacturing costs, integration, etc.

Stepping into more refined details, the TRM sub-block named PCU comprises variables PAt and PhS to adjust and delay the RF signal, respectively,

Figure 3.8: Schematic of an antenna system exploiting mMIMO and advanced beamforming [108].

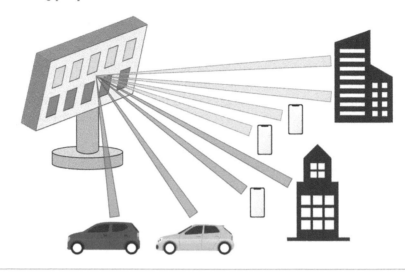

Figure 3.9: Block diagram of a mMIMO system, with TRM and their sub-blocks (PCU and RFFE) [108].

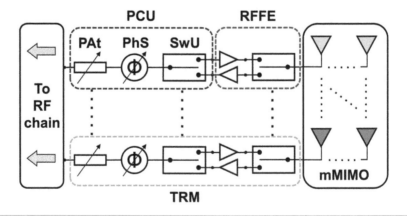

and an SwU, to select the Tx/Rx mode of each single antenna, as shown in the block diagram in Figure 3.9.

RF-MEMS technology (earth) has been already demonstrated in the realization of high-performance, wideband and largely reconfigurable PAt, PhS and SwU. The latter features, in conjunction with high-capacity of the technology in terms of integration, could lead to the monolithic realization of an entire PCU within a unique silicon chip, elevating RF-MEMS from earth to water HW solutions, as mentioned in [108].

3.3.8 Micro/nano computing-in-memory (CIM) devices

Focusing now on a challenge of 6G/FN other than transmission of data, huge demands will arise in terms of pervasive scattering and decentralization (at the edge) of computational and memory capacities (see Chapter 2). As already discussed, the current centralization (to the cloud) of network intelligence is no longer the most efficient strategy, nor is it sustainable in terms of energy consumption and carbon footprint.

Given this frame of reference, it is worth highlighting that critical constraints in terms of speed are not intrinsic to the levels of memory, but inherent to the fact that data must be transferred, via a bus, among physically separated devices, as discussed in [109] and shown in Figure 3.10.

Given these limitations, the integration of data storage and elaboration capacities within unique pieces of HW, known as CIM, can be a viable solution [110–112]. Recalling the discussion developed across the previous parts of this work, micro- and nanotechnologies, and in particular water-like

Figure 3.10: Schematic of the typical stack of CMOS memory devices, according to speed/cost and density/capacity, highlighting the bottlenecks in terms of speed [109].

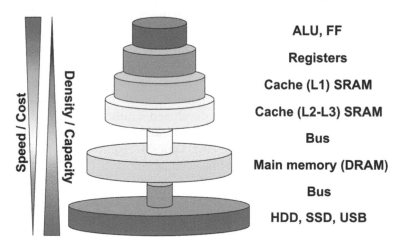

devices, can provide relevant contribution in addressing these challenges and implementing CIM.

Also, interestingly, the scientific community is active in investigating unprecedented approaches to computation, stepping beyond the cornerstone of "0" and "1" digital states, inspired by human brain functioning. Analogue computation principles, known as soft computing, can be powered by micro- and nanotechnologies and materials. To this end, a neuromorphic system called a receptron, reported in [113], opens up complex networks of reconfigurable elements.

In detail, nano-structured gold films are deposited, and their non-linear and non-local electric conduction properties are exploited. In this way, the receptron is able to generate a complete set of Boolean functions of n variables. It also allows the classification of non-linearly separable functions without previous training of the device.

3.4 Summary

This last chapter closed the loop for what concerns the reconceptualization of HW. First, the emerging limitations arising from classical approaches in the development of HW–SW systems were framed by the HW–SW divide concept. A way through was then suggested by increasing the separation and symmetry existing between HW and SW. The core of the discussion was then developed around a conceptual framework, named WEAF Mnecosystem (water, earth, air and fire micro/nanotechnologies ecosystem). Within it, earth and air represent the classical concepts of HW and SW, respectively. Diversely, water addressed simple devices, like sensors and transduces, with diversified functionalities and abilities, among which stand those of self-reacting and self-repairing. Finally, fire addressed micro- and nano-devices dealing with energy, i.e., they are able to harvest energy from the environment, store and transfer it, thus providing power where needed, when needed, in all the tiniest ramifications of the network edge. The chapter also developed a rather extensive discussion of examples in the literature that report MEMS/NEMS devices, technologies and solutions, complying, at least in part, with the requirements and features formally introduced by the WEAF Mnecosystem.

References

[1] P. K. Bondyopadhyay, "Moore's law governs the silicon revolution," *Proc. IEEE*, vol. 86, no. 1, 1998, doi: 10.1109/5.658761

[2]   S. Adee, "The data: 37 years of Moore's law," *IEEE Spectr.*, vol. 45, no. 5, 2008, doi: 10.1109/MSPEC.2008.4505312

[3]   K. J. Kuhn, "Moore's law past 32nm: Future challenges in device scaling," 2009, doi: 10.1109/IWCE.2009.5091124

[4]   G. Strawn and C. Strawn, "Moore's Law at Fifty," *IT Prof.*, vol. 17, no. 6, 2015, doi: 10.1109/MITP.2015.109

[5]   A. B. Kahng, "Scaling: More than Moore's law," *IEEE Des. Test Comput.*, vol. 27, no. 3, 2010, doi: 10.1109/MDT.2010.71

[6]   K. H. Brown, "Moore's law-is there more?," 2000, doi: 10.1109/IMNC.2000.872595

[7]   S. Borkar, "Obeying Moore's law beyond 0.18 micron [microprocessor design]," 2002, doi: 10.1109/asic.2000.880670

[8]   X. Lu, "Beyond Moore's law [in photonic technology]," 2000, doi: 10.1038/nmat1444

[9]   M. Gad-El-Hak, *MEMS: Introduction and fundamentals.* 2005

[10]   M. Gad-El-Hak, *MEMS – Design and Fabrication.* 2005

[11]   M. Gad-El-Hak, *MEMS – Applications.* 2006

[12]   M. Nasrollahzadeh, Z. Issaabadi, M. Sajjadi, S. M. Sajadi, and M. Atarod, "Chapter 2 - Types of Nanostructures," in *An Introduction to Green Nanotechnology*, vol. 28, 2019

[13]   Y. Gogotsi, *Nanomaterials handbook, Second edition.* 2017

[14]   H. Cerjak, "Nanomaterials: an introduction to synthesis, properties and applications," *Mater. Technol.*, vol. 24, no. 2, 2009, doi: 10.1179/175355509x454827

[15]   C. Lamberti and G. Agostini, *Characterization of semiconductor heterostructures and nanostructures.* 2013

[16]   B. K. Kaushik, *Nanoelectronics. Devices, Circuits and Systems – A volume in Advanced Nanomaterials.* 2018

[17]   J. Tian, S. Sosin, J. Iannacci, R. Gaddi, and M. Bartek, "RF–MEMS wafer-level packaging using through-wafer interconnect," *Sensors Actuators A Phys.*, vol. 142, no. 1, pp. 442–451, Mar. 2008, doi: 10.1016/j.sna.2007.09.004

[18]   G. R. Blackwell, *The Electronic Packaging Handbook.* 1999

[19]   A. O. Watanabe, M. Ali, S. Y. B. Sayeed, R. R. Tummala, and P. M. Raj, "A Review of 5G Front-End Systems Package Integration," *IEEE Trans. Components, Packag. Manuf. Technol.*, 2020, doi: 10.1109/TCPMT.2020.3041412

[20]   W. Tian, X. Wang, J. Niu, H. Cui, Y. Chen, and Y. Zhang, "Research status of wafer level packaging for RF MEMS switches," 2020, doi: 10.1109/ICEPT50128.2020.9202925

[21]   G. Chen *et al.*, "Anodic bondable Li-Na-Al-B-Si-O glass-ceramics for Si - ULTCC heterogeneous integration," *J. Eur. Ceram. Soc.*, vol. 39, no. 7, 2019, doi: 10.1016/j.jeurceramsoc.2019.02.028

[22]   G. Murillo, J. Agusti, M. López-Suárez, and A. Gabriel, "Heterogeneous integration of autonomous systems in package for wireless sensor networks," in *Procedia Engineering*, 2011, vol. 25, doi: 10.1016/j.proeng.2011.12.022

[23]   M. Tilli, T. Motooka, V. M. Airaksinen, S. Franssila, M. Paulasto-Kröckel, and V. Lindroos, *Handbook of Silicon Based MEMS Materials and Technologies: Second Edition*. 2015

[24]   *More than Moore or More Moore: a SWOT analysis*. Accessed: Jun. 22, 2023. [Online]. Available: https://tsapps.nist.gov/publication/get_pdf.cfm?pub_id=909225

[25]   D. Kuck, "Keynote talk: A comprehensive approach to HW/SW codesign," 2013, doi: 10.1109/PACT.2013.6618796

[26]   J. Noguera and R. M. Badia, "HW/SW codesign techniques for dynamically reconfigurable architectures," *IEEE Trans. Very Large Scale Integr. Syst.*, vol. 10, no. 4, 2002, doi: 10.1109/TVLSI.2002.801575

[27]   R. Yeniceri and Y. Huner, "HW/SW Codesign and Implementation of an IMU Navigation Filter on Zynq SoC with Linux," 2020, doi: 10.1109/ICEEE49618.2020.9102597

[28]   C. Wolff, I. Gorrochategui, and M. Bücker, "Managing large HW/SW Codesign projects," in *Proceedings of the 6th IEEE International Conference on Intelligent Data Acquisition and Advanced Computing Systems: Technology and Applications, IDAACS'2011*, 2011, vol. 2, doi: 10.1109/IDAACS.2011.6072907

[29]   L. Zhang, Y. C. Liang, and D. Niyato, "6G Visions: Mobile ultra-broadband, super internet-of-things, and artificial intelligence," *China Commun.*, vol. 16, no. 8, 2019, doi: 10.23919/JCC.2019.08.001

[30]   M. Markowski, P. Ryba, and K. Puchala, "Software defined networking research laboratory-experimental topologies and scenarios," 2016, doi: 10.1109/ENIC.2016.044

[31]   T. Theodorou and L. Mamatas, "CORAL-SDN: A software-defined networking solution for the internet of things," in *2017 IEEE Conference on Network Function Virtualization and Software Defined Networks, NFV-SDN 2017*, 2017, vol. 2017-January, doi: 10.1109/NFV-SDN.2017.8169870

[32]   M. H. Dahir, H. Alizadeh, and D. Gözüpek, "Energy efficient virtual network embedding for federated software-defined networks," *Int. J. Commun. Syst.*, vol. 32, no. 6, 2019, doi: 10.1002/dac.3912

[33]   S. O. Mahmudov, "Software Defined Networking: Management of network resources and data flow," 2016, doi: 10.1109/ICISCT.2016.7777384

[34]   A. Al-Quzweeni, T. E. H. El-Gorashi, L. Nonde, and J. M. H. Elmirghani, "Energy efficient network function virtualization in 5G networks," in

International Conference on Transparent Optical Networks, 2015, vol. 2015-August, doi: 10.1109/ICTON.2015.7193559

[35] R. Casellas, R. Vilalta, R. Martinez, and R. Munoz, "Highly available SDN control of flexi-grid networks with network function virtualization-enabled replication," *J. Opt. Commun. Netw.*, vol. 9, no. 2, 2017, doi: 10.1364/JOCN.9.00A207

[36] O. Krasko, H. Al-Zayadi, V. Pashkevych, H. Kopets, and B. Humeniuk, "Network functions virtualization for flexible deployment of converged optical-wireless access infrastructure," in *14th International Conference on Advanced Trends in Radioelectronics, Telecommunications and Computer Engineering, TCSET 2018 - Proceedings*, 2018, vol. 2018-April, doi: 10.1109/TCSET.2018.8336394

[37] W. Guan, X. Wen, L. Wang, Z. Lu, and Y. Shen, "A service-oriented deployment policy of end-to-end network slicing based on complex network theory," *IEEE Access*, vol. 6, 2018, doi: 10.1109/ACCESS.2018.2822398

[38] J. J. A. Esteves, A. Boubendir, F. Guillemin, and P. Sens, "Edge-enabled Optimized Network Slicing in Large Scale Networks," 2020, doi: 10.1109/NoF50125.2020.9249208

[39] W. Wang, W. Guo, and W. Hu, "Network Service Slicing Supporting Ubiquitous Access in Passive Optical Networks," in *International Conference on Transparent Optical Networks*, 2018, vol. 2018-July, doi: 10.1109/ICTON.2018.8473977

[40] K. B. Letaief, W. Chen, Y. Shi, J. Zhang, and Y. J. A. Zhang, "The Roadmap to 6G: AI Empowered Wireless Networks," *IEEE Commun. Mag.*, vol. 57, no. 8, 2019, doi: 10.1109/MCOM.2019.1900271

[41] J. Iannacci and H. V. Poor, "Review and Perspectives of Micro/Nano Technologies as Key-Enablers of 6G," *IEEE Access*, vol. 10, 2022, doi: 10.1109/ACCESS.2022.3176348

[42] J. L. Benson, *Greek Color Theory and the Four Elements*. 2000

[43] J. O'Donohue, *Four Elements: Reflections on Nature*. 2010

[44] R. Frei, R. McWilliam, B. Derrick, A. Purvis, A. Tiwari, and G. Di Marzo Serugendo, "Self-healing and self-repairing technologies," *Int. J. Adv. Manuf. Technol.*, vol. 69, no. 5–8, 2013, doi: 10.1007/s00170-013-5070-2

[45] A. Yao and T. Hikihara, "Reprogrammable logic-memory device of a mechanical resonator," *Int. J. Non. Linear. Mech.*, vol. 94, 2017, doi: 10.1016/j.ijnonlinmec.2016.11.011

[46] F. Y. Kuo, C. Y. Lin, P. C. Chuang, C. L. Chien, Y. L. Yeh, and S. K. A. Wen, "Monolithic Multi-Sensor Design with Resonator-Based MEMS Structures," *IEEE J. Electron Devices Soc.*, vol. 5, no. 3, 2017, doi: 10.1109/JEDS.2017.2666821

[47]   M. Hautefeuille, B. O'Flynn, F. Peters, and C. O'Mahony, "Miniaturised multi-MEMS sensor development," *Microelectron. Reliab.*, vol. 49, no. 6, 2009, doi: 10.1016/j.microrel.2009.02.017

[48]   Q. Ma, Z. Wang, and L. Pan, "Monolithic integration of multiple sensors on a single silicon chip," 2016, doi: 10.1109/DTIP.2016.7514831

[49]   R. Seto *et al.*, "Development and use of a micro optical blood flow sensor based on system in package (SiP) technology that fuses optical MEMS and integrated circuit to detect avian influenza," 2007, doi: 10.1109/SENSOR.2007.4300543

[50]   C. S. Premachandran *et al.*, "A novel, wafer-level stacking method for low-chip yield and non-uniform, chip-size wafers for MEMS and 3D SIP applications," 2008, doi: 10.1109/ECTC.2008.4549988

[51]   T. Bieniek *et al.*, "Novel methodology for 3D MEMS-IC design and Co-simulation on MEMS microphone smart system example," 2014, doi: 10.1109/3DIC.2014.7152184

[52]   J. Iannacci, "Microsystem based Energy Harvesting (EH-MEMS): Powering pervasivity of the Internet of Things (IoT) - A review with focus on mechanical vibrations," *J. King Saud Univ. - Sci.*, 2017, doi: 10.1016/j.jksus.2017.05.019

[53]   J. Yan, X. Liao, S. Ji, and S. Zhang, "A Novel Multi-Source Micro Power Generator for Harvesting Thermal and Optical Energy," *IEEE Electron Device Lett.*, vol. 40, no. 2, 2019, doi: 10.1109/LED.2018.2889300

[54]   T. Wang *et al.*, "Self-healing and flexible carbon nanotube/polyurethane composite for efficient electromagnetic interference shielding," *Compos. Part B Eng.*, vol. 193, 2020, doi: 10.1016/j.compositesb.2020.108015

[55]   C. Lesaint *et al.*, "Self-healing high voltage electrical insulation materials," 2014, doi: 10.1109/EIC.2014.6869384

[56]   L. Q. Zhou *et al.*, "Ultrasonic Inspection and Self-Healing of Ge and 3C-SiC Semiconductor Membranes," *J. Microelectromechanical Syst.*, 2020, doi: 10.1109/JMEMS.2020.2981909

[57]   H. Bai, Y. S. Kim, and R. F. Shepherd, "Autonomous self-healing optical sensors for damage intelligent soft-bodied systems," *Sci. Adv.*, vol. 8, no. 49, 2022, doi: 10.1126/sciadv.abq2104

[58]   M. Farnsworth and A. Tiwari, "Modelling, Simulation and Analysis of a Self-healing Energy Harvester," in *Procedia CIRP*, 2015, vol. 38, doi: 10.1016/j.procir.2015.07.084

[59]   S. Ilyas, S. Ahmed, M. A. A. Hafiz, H. Fariborzi, and M. I. Younis, "Cascadable microelectromechanical resonator logic gate," *J. Micromechanics Microengineering*, vol. 29, no. 1, 2019, doi: 10.1088/1361-6439/aaf0e6

[60] W. W. Jang *et al.*, "NEMS switch with 30 nm-thick beam and 20 nm-thick air-gap for high density non-volatile memory applications," *Solid. State. Electron.*, vol. 52, no. 10, 2008, doi: 10.1016/j.sse.2008.06.026

[61] J. Iannacci, "RF-MEMS technology as an enabler of 5G: Low-loss ohmic switch tested up to 110 GHz," *Sensors Actuators, A Phys.*, vol. 279, 2018, doi: 10.1016/j.sna.2018.07.005

[62] J. Iannacci, *RF-MEMS Technology for High-Performance Passives – 5G applications and prospects for 6G*, 2nd ed. 2022

[63] J. Iannacci, "RF-MEMS for high-performance and widely reconfigurable passive components - A review with focus on future telecommunications, Internet of Things (IoT) and 5G applications," *J. King Saud Univ. - Sci.*, 2017, doi: 10.1016/j.jksus.2017.06.011

[64] J. Iannacci, M. Huhn, C. Tschoban, and H. Pötter, "RF-MEMS Technology for Future (5G) Mobile and High-Frequency Applications: Reconfigurable 8-Bit Power Attenuator Tested up to 110 GHz," *IEEE Electron Device Lett.*, vol. 37, no. 12, 2016, doi: 10.1109/LED.2016.2623328

[65] F. Giacomozzi, V. Mulloni, S. Colpo, J. Iannacci, B. Margesin, and A. Faes, "A flexible fabrication process for RF MEMS devices," *Rom. J. Inf. Sci. Technol.*, vol. 14, no. 3, 2011

[66] *Status of the MEMS Industry 2021*. Accessed: July. 5, 2023. [Online]. Available: https://s3.i-micronews.com/uploads/2021/07/YINTR21180-Status-of-the-MEMS-Industry-2021_Flyer.pdf

[67] A. Goritz *et al.*, "BEOL modifications of a 130 nm SiGe BiCMOS technology for monolithic integration of thin-film wafer-level encapsulated D-Band RF-MEMS switches," 2021, doi: 10.1109/DTIP54218.2021.9568672

[68] N. Zhang, R. Song, J. Liu, and J. Yang, "A Packaged THz Shunt RF MEMS Switch with Low Insertion Loss," *IEEE Sens. J.*, 2021, doi: 10.1109/JSEN.2021.3113647

[69] T. J. Cui, D. R. Smith, and R. Liu, *Metamaterials: Theory, design, and applications*. 2010

[70] T. Sun *et al.*, "Terahertz Beam Steering based on CMOS Tunable Metamaterials," in *International Conference on Infrared, Millimeter, and Terahertz Waves, IRMMW-THz*, 2021, vol. 2021-August, doi: 10.1109/IRMMW-THz50926.2021.9567412

[71] Z. N. Chen, T. Li, and W. E. I. Liu, "Microwave Metasurface-based Lens Antennas for 5G and beyond," 2020, doi: 10.23919/EuCAP48036.2020.9135285

[72] M. Jiang, Z. N. Chen, Y. Zhang, W. Hong, and X. Xuan, "Metamaterial-Based Thin Planar Lens Antenna for Spatial Beamforming and

Multibeam Massive MIMO," *IEEE Trans. Antennas Propag.*, vol. 65, no. 2, 2017, doi: 10.1109/TAP.2016.2631589

[73] V. K. Khanna, *Flexible Electronics, Volume 1*. 2019

[74] V. K. Khanna, *Flexible Electronics, Volume 2*. 2019

[75] V. K. Khanna, *Flexible Electronics, Volume 3*. 2019

[76] K. Takei, "Printed Multifunctional Flexible Healthcare Patch," 2018, doi: 10.1109/IFETC.2018.8583894

[77] S. Tokito, "Flexible Printed Organic Thin-Film Transistor Devices and IoT Sensor Applications," 2018, doi: 10.1109/EDTM.2018.8421492

[78] J. Ethier, R. Chaharmir, J. Shaker, and K. Hettak, "Electromagnetic Engineered Surface Gratings at 5G Bands Using Printed Electronics," 2018, doi: 10.1109/IFETC.2018.8583977

[79] P. Ashbum *et al.*, "Si1-x,Gex Heterojunction Bipolar Transistors: the future of silicon bipolar technology or not?," Proc. of ESSDERC '93, 1993, doi: n/a

[80] M. Sathishkumar, T. S. Arun Samuel, and P. Vimala, "A Detailed Review on Heterojunction Tunnel Field Effect Transistors," 2020, doi: 10.1109/ic-ETITE47903.2020.197

[81] A. Di Bartolomeo, "Graphene Schottky diodes: An experimental review of the rectifying graphene/semiconductor heterojunction," *Physics Reports*, vol. 606. 2016, doi: 10.1016/j.physrep.2015.10.003

[82] B. Asllani, H. Morel, P. Bevilacqua, and D. Planson, "Demonstration of the Short-circuit Ruggedness of a 10 kV Silicon Carbide Bipolar Junction Transistor," 2020, doi: 10.23919/EPE20ECCEEurope43536.2020.9215769

[83] N. Kashio *et al.*, "Improvement of High-Frequency Characteristics of InGaAsSb-Base Double Heterojunction Bipolar Transistors by Inserting a Highly Doped GaAsSb Base Contact Layer," *IEEE Electron Device Letters*, vol. 36, no. 7, 2015, doi: 10.1109/LED.2015.2429142

[84] M. Schroter and A. Pawlak, "SiGe heterojunction bipolar transistor technology for sub-mm-wave electronics - State-of-the-art and future prospects," in *SIRF 2018 - 2018 IEEE 18th Topical Meeting on Silicon Monolithic Integrated Circuits in RF Systems*, 2018, vol. 2018-January, doi: 10.1109/SIRF.2018.8304230

[85] R. Narzary, P. Phukan, and P. P. Sahu, "Efficiency Enhancement of Low-Cost Heterojunction Solar Cell by the Incorporation of Highly Conducting rGO into ZnO Nanostructure," *IEEE Trans. Electron Devices*, vol. 68, no. 7, 2021, doi: 10.1109/TED.2021.3080228

[86] R. S. Pal, S. Sharma and S. Dasgupta, "Recent trend of FinFET devices and its challenges: A review," 2017, doi: 10.1109/ICEDSS.2017.8073675

[87] M. C. Hsieh, S. Lin, I. Hsu, C. Y. Chen, and N. J. Cho, "Fine pitch high bandwidth flip chip package-on-package development," in *EMPC 2017 -*

21st European Microelectronics and Packaging Conference and Exhibition, 2018, vol. 2018-January, doi: 10.23919/EMPC.2017.8346847

[88] T. N. Chang, C. Y. Tsou, B. H. Wang, and K. N. Chiang, "Novel wafer level packaging for large die size device," 2017, doi: 10.23919/ICEP.2017.7939378

[89] S. Li, *SiP-System in Package Design and Simulation.* 2017

[90] K. Lee, "High-Density Fan-Out Technology for Advanced SiP and 3D Heterogeneous Integration," 2018, doi: 10.1109/IRPS.2018.8353588

[91] L. Gargalis *et al.,* "Additive manufacturing and testing of a soft magnetic rotor for a switched reluctance motor," *IEEE Access,* vol. 8, 2020, doi: 10.1109/ACCESS.2020.3037190

[92] S. Zhang, W. Whittow, D. Cadman, R. Mittra, and Jy. C. Vardaxoglou, "Additive Manufacturing for High Performance Antennas and RF Components," 2019, doi: 10.1109/IEEE-IWS.2019.8803912

[93] M. Szymkiewicz, Y. Konkel, C. Hartwanger, and M. Schneider, "Ku-band sidearm orthomode transducer manufactured by additive layer manufacturing," 2016, doi: 10.1109/EuCAP.2016.7481434

[94] S. Zhang *et al.,* "3D Antennas, Metamaterials, and Additive Manufacturing," 2019, doi: 10.1109/IEEE-IWS.2019.8803909

[95] J. Wang, C. Shao, Y. Wang, L. Sun, and Y. Zhao, "Microfluidics for Medical Additive Manufacturing," *Engineering,* vol. 6, no. 11, 2020, doi: 10.1016/j.eng.2020.10.001

[96] S. Zhang, W. Whittow, D. Cadman, R. Mittra, and Jy. C. Vardaxoglou, "Additive Manufacturing for High Performance Antennas and RF Components," 2019, doi: 10.1109/IEEE-IWS.2019.8803912

[97] T. Takeshita, T. Yamashita, N. Makimoto, and T. Kobayashi, "Development of ultra-thin MEMS micro mirror device," 2017, doi: 10.1109/TRANSDUCERS.2017.7994499

[98] R. Blue, L. Li, G. M. H. Flockhart, and D. Uttamchandani, "Wavelength filtering using MEMS actuated coupling from optical fibres to spherical resonators," 2011, doi: 10.1109/OMEMS.2011.6031024

[99] H. Dessalegn and T. Srinivas, "Optical MEMS pressure sensor based on double ring resonator," 2013, doi: 10.1109/ICMAP.2013.6733554

[100] K. Vahala, "New Directions for High-Q Optical Micro-Resonators: Soliton-Based Optical Clocks to Compact Sagnac Gyros," 2016, doi: 10.1109/OMN.2016.7565874

[101] G. N. Nielson *et al.,* "Integrated wavelength-selective optical MEMS switching using ring resonator filters," *IEEE Photonics Technol. Lett.,* vol. 17, no. 6, 2005, doi: 10.1109/LPT.2005.846951

[102] Z. Da Li *et al.,* "Experimental demonstration of all-photonic quantum repeater," 2019, doi: 10.23919/CLEO.2019.8750288

[103]   J. W. Silverstone *et al.*, "Silicon quantum photonics in the short-wave infrared: A new platform for big quantum optics," 2019, doi: 10.1109/CLEOE-EQEC.2019.8871561

[104]   Q. Liu and M. P. Fok, "Real-Time RF Multi-Dimensional Signal Switching Using Polarization-Dependent Optical Mixing," *IEEE Photonics J.*, vol. 12, no. 2, 2020, doi: 10.1109/JPHOT.2020.2977847

[105]   A. L. M. Muniz, R. M. Borges, R. N. Da Silva, D. F. Noque, and S. Arismar Cerqueira, "Ultra-broadband photonics-based RF front-end toward 5G networks," *J. Opt. Commun. Netw.*, vol. 8, no. 11, 2016, doi: 10.1364/JOCN.8.000B35

[106]   D. Heinz *et al.*, "GaInN Quantum Wells as Optochemical Transducers for Chemical Sensors and Biosensors," *IEEE J. Sel. Top. Quantum Electron.*, vol. 23, no. 2, 2017, doi: 10.1109/JSTQE.2016.2617818

[107]   W. Jiang and F.-L. Luo, *6G Key Technologies: A Comprehensive Guide*, 2023, doi: 10.1002/9781119847502

[108]   J. Iannacci, "A Perspective Vision of Micro/Nano Systems and Technologies as Enablers of 6G, Super-IoT, and Tactile Internet," *Proc. IEEE*, vol. 111, no. 1, 2023, doi: 10.1109/JPROC.2022.3223791

[109]   W. Kang, H. Zhang, and W. Zhao, "Spintronic Memories: From Memory to Computing-in-Memory," 2019, doi: 10.1109/NANOARCH47378.2019.181298

[110]   C. X. Xue *et al.*, "24.1 A 1Mb Multibit ReRAM Computing-In-Memory Macro with 14.6ns Parallel MAC Computing Time for CNN Based AI Edge Processors," in *Digest of Technical Papers - IEEE International Solid-State Circuits Conference*, 2019, vol. 2019-February, doi: 10.1109/ISSCC.2019.8662395

[111]   S. Jain, A. Ranjan, K. Roy, and A. Raghunathan, "Computing in Memory with Spin-Transfer Torque Magnetic RAM," *IEEE Trans. Very Large Scale Integr. Syst.*, vol. 26, no. 3, 2017, doi: 10.1109/TVLSI.2017.2776954

[112]   J. Yue *et al.*, "A 65nm Computing-in-Memory-Based CNN Processor with 2.9-to-35.8TOPS/W System Energy Efficiency Using Dynamic-Sparsity Performance-Scaling Architecture and Energy-Efficient Inter/Intra-Macro Data Reuse," in *Digest of Technical Papers - IEEE International Solid-State Circuits Conference*, 2020, vol. 2020-February, doi: 10.1109/ISSCC19947.2020.9062958

[113]   G. Martini, M. Mirigliano, B. Paroli, and P. Milani, "The Receptron: A device for the implementation of information processing systems based on complex nanostructured systems," *Japanese Journal of Applied Physics*, vol. 61, no. SM. 2022, doi: 10.35848/1347-4065/ac665c

Index

Milton Keynes UK
Ingram Content Group UK Ltd.
UKHW022041141024
449569UK00014B/680

9 788770 040792